Glencoe McGraw-Hill

Math Connects

Course 1

Study Guide and Intervention and Practice Workbook

To the Student This *Study Guide and Intervention and Practice Workbook* gives you additional examples and problems for the concept exercises in each lesson. The exercises are designed to aid your study of mathematics by reinforcing important mathematical skills needed to succeed in the everyday world. The materials are organized by chapter and lesson, with one *Study Guide and Intervention and Practice* worksheet for every lesson in *Glencoe Math Connects, Course 1*.

Always keep your workbook handy. Along with your textbook, daily homework, and class notes, the completed *Study Guide and Intervention and Practice Workbook* can help you in reviewing for quizzes and tests.

To the Teacher These worksheets are the same ones found in the Chapter Resource Masters for *Glencoe Math Connects, Course 1*. The answers to these worksheets are available at the end of each Chapter Resource Masters booklet as well as in your Teacher Wraparound Edition interleaf pages.

The McGraw-Hill Companies

Copyright © by The McGraw-Hill Companies, Inc. All rights reserved.
Except as permitted under the United States Copyright Act, no part of this publication may be reproduced or distributed in any form or by any means, or stored in a database or retrieval system, without prior written permission of the publisher.

Send all inquiries to:
Glencoe/McGraw-Hill
8787 Orion Place
Columbus, OH 43240

ISBN: 978-0-07-881032-9
MHID: 0-07-881032-9

Study Guide and Intervention and Practice Workbook, Course 1

Printed in the United States of America
11 12 13 14 15 16 17 18 19 RHR 13 12 11

CONTENTS

Lesson/Title		Page
11-6	Dividing Integers	179
11-6	Dividing Integers	180
11-7	The Coordinate Plane	181
11-7	The Coordinate Plane	182
11-8	Translations	183
11-8	Translations	184
11-9	Reflections	185
11-9	Reflections	186
11-10	Rotations	187
11-10	Rotations	188
12-1	The Distributive Property	189
12-1	The Distributive Property	190
12-2	Simplifying Algebraic Expressions	191
12-2	Simplifying Algebraic Expressions	192
12-3	Solving Addition Equations	193
12-3	Solving Addition Equations	194
12-4	Solving Subtraction Equations	195
12-4	Solving Subtraction Equations	196
12-5	Solving Multiplication Equations	197
12-5	Solving Multiplication Equations	198
12-6	Problem-Solving Investigation: Choose the Best Method of Computation	199
12-6	Problem-Solving Investigation: Choose the Best Method of Computation	200

NAME ______________________ DATE ____________ PERIOD _____

1-1 Study Guide and Intervention

A Plan for Problem Solving

When solving problems, it is helpful to have an organized plan to solve the problem.
The following four steps can be used to solve any math problem.

1 Understand – Read and get a general understanding of the problem.

2 Plan – Make a plan to solve the problem and estimate the solution.

3 Solve – Use your plan to solve the problem.

4 Check – Check the reasonableness of your solution.

Example 1 SPORTS **The table shows the number of field goals made by Henry High School's top three basketball team members during last year's season. How many more field goals did Brad make than Denny?**

Name	3-Point Field Goals
Brad	216
Chris	201
Denny	195

Understand You know the number of field goals made. You need to find how many more field goals Brad made than Denny.

Plan Use only the needed information, the goals made by Brad and Denny. To find the difference, subtract 195 from 216.

Solve 216 − 195 = 21; Brad made 21 more field goals than Denny.

Check Check the answer by adding. Since 195 + 21 = 216, the answer is correct.

Exercises

1. During which step do you check your work to make sure your answer is correct?

2. Explain what you do during the first step of the problem-solving plan.

SPORTS For Exercises 3 and 4, use the field goal table above and the four-step plan.

3. How many more field goals did Chris make than Denny?

4. How many field goals did the three boys make all together?

Copyright © Glencoe/McGraw-Hill, a division of The McGraw-Hill Companies, Inc.

NAME ______________________________ DATE ____________ PERIOD _____

1-1 Practice

A Plan for Problem Solving

PATTERNS Complete each pattern.

1. 17, 21, 25, 29, _____, _____, _____,

2. 32, 29, 26, 23, _____, _____, _____,

3. 1, 2, 4, 7, _____, _____, _____,

4. 64, 32, 16, 8, _____, _____, _____,

5. **ANALYZE GRAPHS** Refer to the graph. How many acres smaller is Lake Meredith National Recreation Area than Big Thicket National Preserve?

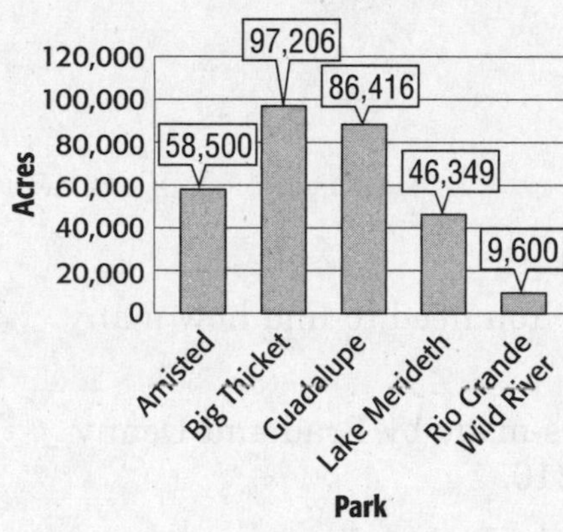

6. **TRAVEL** The distance between Dallas and Beaumont is about 290 miles. Henry drove from Dallas to Beaumont at 58 miles per hour. How many hours did it take Henry to reach Beaumont?

7. **ANALYZE TABLES** The table lists the times that ferries leave the terminal every day. At what times will the next three ferries leave the terminal?

6:36 A.M.
7:11 A.M.
7:17 A.M.
7:52 A.M.
7:58 A.M.

8. **MONEY** The Wilsons bought a refrigerator and a stove for a total cost of $745. They will pay for the purchase in five equal payments. What will be the amount of each payment?

9. **MUSIC** Luanda practices playing the piano for 24 minutes each day. How many hours does she practice in one year?

Copyright © Glencoe/McGraw-Hill, a division of The McGraw-Hill Companies, Inc.

NAME _______________ DATE _______ PERIOD _____

1-2 Study Guide and Intervention

Prime Factors

Factors are the numbers that are multiplied to get a product. A product is the answer to a multiplication problem. A **prime number** is a whole number that has only 2 factors, 1 and the number itself. A **composite number** is a number greater than 1 with more than two factors.

Example 1 **Tell whether each number is *prime*, *composite*, or *neither*.**

Number	Factors	Prime or Composite?
15	1×15 3×5	Composite
17	1×17	Prime
1	1	Neither

Example 2 **Find the prime factorization of 18.**

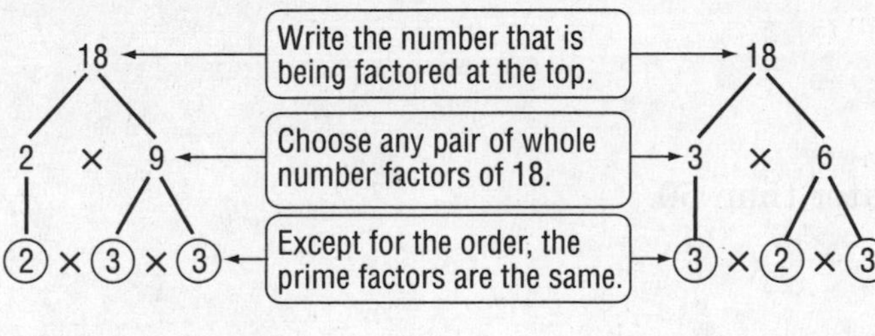

18 is divisible by 2, because the ones digit is divisible by 2.

Circle the prime number, 2.

9 is divisible by 3, because the sum of the digits is divisible by 3.

Circle the prime numbers, 3 and 3.

The prime factorization of 18 is $2 \times 3 \times 3$.

Exercises

Tell whether each number is *prime*, *composite*, or *neither*.

1. 7 **2.** 12 **3.** 29

4. 81 **5.** 18 **6.** 23

7. 54 **8.** 28 **9.** 120

10. 243 **11.** 61 **12.** 114

Find the prime factorization of each number.

13. 125 **14.** 44

15. 11 **16.** 56

Copyright © Glencoe/McGraw-Hill, a division of The McGraw-Hill Companies, Inc.

Lesson 1-2

NAME ______________________________ DATE ____________ PERIOD ____

1-2 Practice

Prime Factors

Tell whether each number is *prime, composite,* or *neither*.

1. 24 **2.** 1 **3.** 13 **4.** 25

5. 91 **6.** 0 **7.** 181 **8.** 145

Find the prime factorization of each number.

9. 16 **10.** 48 **11.** 66

12. 56 **13.** 80 **14.** 95

15. Find the least prime number that is greater than 50.

16. All odd numbers greater than 7 can be expressed as the sum of three prime numbers. Which three prime numbers have a sum of 43? Justify your answer.

17. **GARDENING** Julia wants to plant 24 tomato plants in rows. Each row will have the same number of plants in it. Find three possible numbers of rows and the number of plants in each row.

18. **SHOPPING** Jamal bought boxes of nails that each cost the same. He spent a total of $42. Find three possible costs per box and the number of boxes that he could have purchased.

Copyright © Glencoe/McGraw-Hill, a division of The McGraw-Hill Companies, Inc.

NAME ______________________ DATE ____________ PERIOD _____

1-3 Study Guide and Intervention

Powers and Exponents

A product of prime factors can be written using exponents and a base. Numbers expressed using exponents are called **powers**.

Powers	Words	Expression	Value
4^2	4 to the second power or 4 squared	4×4	16
5^6	5 to the sixth power	$5 \times 5 \times 5 \times 5 \times 5 \times 5$	15,625
7^4	7 to the fourth power	$7 \times 7 \times 7 \times 7$	2,401
9^3	9 to the third power or 9 cubed	$9 \times 9 \times 9$	729

Example 1 **Write $6 \times 6 \times 6$ using an exponent. Then find the value.**

The base is 6. Since 6 is a factor 3 times, the exponent is 3.
$6 \times 6 \times 6 = 6^3$ or 216

Example 2 **Write 2^4 as a product of the same factor. Then find the value.**

The base is 2. The exponent is 4. So, 2 is a factor 4 times.
$2^4 = 2 \times 2 \times 2 \times 2$ or 16

Example 3 **Write the prime factorization of 225 using exponents.**

The prime factorization of 225 can be written as $3 \times 3 \times 5 \times 5$, or $3^2 \times 5^2$.

Exercises

Write each product using an exponent. Then find the value.

1. $2 \times 2 \times 2 \times 2 \times 2$

2. 9×9

3. $3 \times 3 \times 3$

4. $5 \times 5 \times 5$

5. $3 \times 3 \times 3 \times 3 \times 3$

6. 10×10

Write each power as a product of the same factor. Then find the value.

7. 7^2

8. 4^3

9. 8^4

10. 5^5

11. 2^8

12. 7^3

Write the prime factorization of each number using exponents.

13. 40

14. 75

15. 100

16. 147

Lesson 1-3

Copyright © Glencoe/McGraw-Hill, a division of The McGraw-Hill Companies, Inc.

NAME ______________________ DATE ____________ PERIOD _____

1-3 Practice

Powers and Exponents

Write each product using an exponent.

1. 6×6

2. $10 \times 10 \times 10 \times 10$

3. $4 \times 4 \times 4 \times 4 \times 4$

4. $8 \times 8 \times 8 \times 8 \times 8 \times 8 \times 8 \times 8$

5. $5 \times 5 \times 5 \times 5 \times 5 \times 5$

6. $13 \times 13 \times 13$

Write each power as a product of the same factor. Then find the value.

7. 10^1

8. 2^7

9. 8^3

10. 3^8

11. nine squared

12. four to the sixth power

Write the prime factorization of each number using exponents.

13. 32

14. 100

15. 63

16. 99

17. 52

18. 147

19. **LABELS** A sheet of labels has 8 rows of labels with 8 labels in each row. How many total labels are on the sheet? Write your answer using exponents, and then find the value.

20. **CANDLES** To find how much wax the candle mold holds, use the expression $s \times s \times s$, where s is the length of a side. Write this expression as a power. The amount of wax the mold holds is measured in cubic units. How many cubic units of wax does the mold hold?

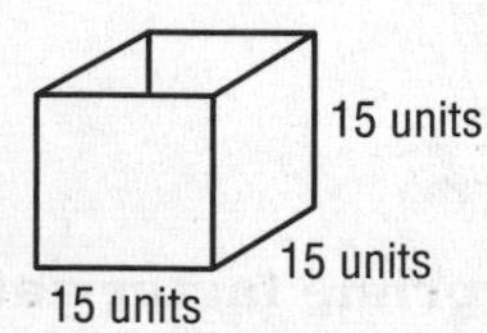

Copyright © Glencoe/McGraw-Hill, a division of The McGraw-Hill Companies, Inc.

NAME ______________________ DATE ____________ PERIOD _____

1-4 Study Guide and Intervention

Order of Operations

Order of Operations

1. Simplify the expressions inside grouping symbols, like parentheses.
2. Find the value of all powers.
3. Multiply and divide in order from left to right.
4. Add and subtract in order from left to right.

Example 1 **Find the value of $48 \div (3 + 3) - 2^2$.**

$48 \div (3 + 3) - 2^2 = 48 \div 6 - 2^2$	Simplify the expression inside the parentheses.
$= 48 \div 6 - 4$	Find 2^2.
$= 8 - 4$	Divide 48 by 6.
$= 4$	Subtract 4 from 8.

Example 2 **Write and solve an expression to find the total cost of planting flowers in the garden.**

Item	Cost Per Item	Number of Items Needed
pack of flowers	$4	5
bag of dirt	$3	1
bottle of fertilizer	$4	1

Words	cost of 5 flower packs	plus	cost of dirt	plus	cost of fertilizer
Expression	5 × $4	+	$3	+	$4

$$5 \times \$4 + \$3 + \$4 = \$20 + \$3 + \$4$$
$$= \$23 + \$4$$
$$= \$27$$

The total cost of planting flowers in the garden is $27.

Exercises

Find the value of each expression.

1. $7 + 2 \times 3$

2. $12 \div 3 + 5$

3. $16 - (4 + 5)$

4. $8 \times 8 \div 4$

5. $10 + 14 \div 2$

6. $3 \times 3 + 2 \times 4$

7. $80 - 8 \times 3^2$

8. $11 \times (9 - 2^2)$

9. $25 \div 5 + 6 \times (12 - 4)$

10. GARDENING Refer to Example 2 above. Suppose that the gardener did not buy enough flowers and goes back to the store to purchase four more packs. She also purchases a hoe for $16. Write an expression that shows the total amount she spent to plant flowers in her garden.

Lesson 1-4

Copyright © Glencoe/McGraw-Hill, a division of The McGraw-Hill Companies, Inc.

NAME ______________________ DATE ____________ PERIOD _____

1-4 Practice

Order of Operations

Find the value of each expression.

1. $34 + 17 - 5$

2. $25 - 14 + 3$

3. $42 + 6 \div 2$

4. $39 \times (15 \div 3) - 16$

5. $48 \div 8 + 5 \times (7 - 2)$

6. $64 \div (15 - 7) \times 2 - 9$

7. $(3 + 7) \times 6 + 4$

8. $9 + 8 \times 3 - (5 \times 2)$

9. $7^2 + 6 \times 2$

10. $34 - 8^2 \div 4$

11. $45 \div 3 \times 2^3$

12. $4 \times (5^2 - 12) - 6$

13. $78 - 2^4 \div (14 - 6) \times 2$

14. $9 + 7 \times (15 + 3) \div 3^2$

15. $13 + (4^3 \div 2) \times 5 - 17$

16. Using symbols, write the product of 18 and 7 plus 5.

ART For Exercises 17 and 18, use the following information.

An art supply store sells posters for \$9 each and picture frames for \$15 each.

17. Write an expression for the total cost of 6 posters and 6 frames.

18. What is the total cost for 6 framed posters?

19. **SCIENCE** There are 24 students in a science class. Mr. Sato will give each pair of students 3 magnets. So far, Mr. Sato has given 9 pairs of students their 3 magnets. How many more magnets does Mr. Sato need so that each pair of students has exactly 3 magnets?

Copyright © Glencoe/McGraw-Hill, a division of The McGraw-Hill Companies, Inc.

NAME ______________________ DATE ____________ PERIOD _____

1-5 Study Guide and Intervention

Algebra: Variables and Expressions

- A **variable** is a symbol, usually a letter, used to represent a number.
- Multiplication in algebra can be shown as $4n$, $4 \cdot n$, or $4 \times n$.
- **Algebraic expressions** are combinations of variables, numbers, and at least one operation.

Example 1 **Evaluate $35 + x$ if $x = 6$.**

$35 + x = 35 + 6$	Replace x with 6.
$= 41$	Add 35 and 6.

Example 2 **Evaluate $y + x$ if $x = 21$ and $y = 35$.**

$y + x = 35 + 21$	Replace x with 21 and y with 35.
$= 56$	Add 35 and 21.

Example 3 **Evaluate $4n + 3$ if $n = 2$.**

$4n + 3 = 4 \times 2 + 3$	Replace n with 2.
$= 8 + 3$	Find the product of 4 and 2.
$= 11$	Add 8 and 3.

Example 4 **Evaluate $4n - 2$ if $n = 5$.**

$4n - 2 = 4 \times 5 - 2$	Replace n with 5.
$= 20 - 2$	Find the product of 4 and 5.
$= 18$	Subtract 2 from 20.

Exercises

Evaluate each expression if $y = 4$.

1. $3 + y$ **2.** $y + 8$ **3.** $4 \times y$

4. $9y$ **5.** $15y$ **6.** $300y$

7. y^2 **8.** $y^2 + 18$ **9.** $y^2 + 3 \times 7$

Evaluate each expression if $m = 3$ and $k = 10$.

10. $16 + m$ **11.** $4k$ **12.** $m \times k$

13. $m + k$ **14.** $7m + k$ **15.** $6k + m$

16. $3k - 4m$ **17.** $2mk$ **18.** $5k - 6m$

19. $20m \div k$ **20.** $m^3 + 2k^2$ **21.** $k^2 \div (2 + m)$

Copyright © Glencoe/McGraw-Hill, a division of The McGraw-Hill Companies, Inc.

NAME ______________________ DATE __________ PERIOD _____

1-5 Practice

Algebra: Variables and Expressions

Evaluate each expression if $m = 6$ and $n = 12$.

1. $m + 5$

2. $n - 7$

3. $m \cdot 4$

4. $m + n$

5. $n - m$

6. $12 \div n$

7. $9 \cdot n$

8. $n \div m$

9. $2m + 5$

10. $4m - 17$

11. $36 - 6m$

12. $3n + 8$

Evaluate each expression if $a = 9$, $b = 3$, and $c = 12$.

13. $4a - 17$

14. $14 + 2c$

15. $c \div 2$

16. ac

17. $c \div b$

18. $2ac$

19. $b^3 + c$

20. $19 + 6a \div 2$

21. $4b^2 \cdot 3$

22. $3c \div (2b^2)$

23. $c^2 - (3a)$

24. $ac \div (2b)$

25. **ANIMALS** A Gentoo penguin can swim at a rate of 17 miles per hour. How many miles can a penguin swim in 4 hours? Use the expression rt, where r represents rate and t represents time.

26. **CLOTHING** A company charges \$6 to make a pattern for an order of T-shirts and \$11 for each T-shirt it produces from the pattern. The expression \$11n + \$6 represents the cost of n T-shirts with the same pattern. Find the total cost for 5 T-shirts with the same pattern.

Copyright © Glencoe/McGraw-Hill, a division of The McGraw-Hill Companies, Inc.

NAME ______________________________ DATE ____________ PERIOD _____

1-6 Study Guide and Intervention

Algebra: Functions

A **function rule** describes the relationship between the input and output of a **function**. The inputs and outputs can be organized in a **function table**.

Example 1 **Complete the function table.**

Input (x)	Output ($x - 3$)
9	■
8	■
6	■

The function rule is $x - 3$. Subtract 3 from each input.

Input		Output
9	$-3 \rightarrow$	6
8	$-3 \rightarrow$	5
6	$-3 \rightarrow$	3

$\rightarrow$

Input (x)	Output ($x - 3$)
9	6
8	5
6	3

Example 2 **Find the rule for the function table.**

Input (x)	Output (■)
0	0
1	4
2	8

Study the relationship between each input and output.

Input		Output
0	$\times 4 \rightarrow$	0
1	$\times 4 \rightarrow$	4
2	$\times 4 \rightarrow$	8

The output is four times the input. So, the function rule is $4x$.

Exercises

Complete each function table.

1.

Input (x)	Output ($2x$)
0	
2	
4	

2.

Input (x)	Output ($4 + x$)
0	
1	
4	

Find the rule for each function table.

3.

Input (x)	Output (■)
1	3
2	4
5	7

4.

Input (x)	Output (■)
2	1
6	3
10	5

Copyright © Glencoe/McGraw-Hill, a division of The McGraw-Hill Companies, Inc.

NAME ______________________ DATE ____________ PERIOD _____

1-6 Practice

Algebra: Functions

Complete each function table.

1.

Input (x)	Output ($x + 6$)
0	
3	
7	

2.

Input (x)	Output ($x - 1$)
1	
4	
8	

3.

Input (x)	Output ($3x$)
0	
2	
4	

4.

Input (x)	Output ($x \div 2$)
4	
8	
10	

Find the rule for each function table.

5.

x	■
4	1
8	2
16	4

6.

x	■
12	8
13	9
15	11

7.

x	■
2	1
6	3
10	5

8.

x	■
3	0
5	2
6	3
8	5
11	8

9.

x	■
0	3
1	6
2	9
3	12
4	15

10.

x	■
2	5
4	13
6	21
8	29
10	37

11. **FOOD** A pizza place sells pizzas for $7 each plus a $4 delivery charge per order. If Pat orders 3 pizzas to be delivered, what will be his total cost?

12. **MOVIES** A store sells used DVDs for $8 each and used videotapes for $6 each. Write a function rule to represent the total selling price of DVDs (d) and videotapes (v). Then use the function rule to find the price of 5 DVDs and 3 videotapes.

Copyright © Glencoe/McGraw-Hill, a division of The McGraw-Hill Companies, Inc.

NAME ______________________ DATE ____________ PERIOD _____

1-7

Study Guide and Intervention

Problem-Solving Investigation: Guess and Check

Lesson 1–7

When solving problems, one strategy that is helpful to use is *guess and check*. Based on the information in the problem, you can make a guess of the solution. Then use computations to check if your guess is correct. You can repeat this process until you find the correct solution.

You can use guess and check, along with the following four-step problem solving plan to solve a problem.

1 Understand – Read and get a general understanding of the problem.

2 Plan – Make a plan to solve the problem and estimate the solution.

3 Solve – Use your plan to solve the problem.

4 Check – Check the reasonableness of your solution.

Example 1 SPORTS **Meagan made a combination of 2-point baskets and 3-point baskets in the basketball game. She scored a total of 9 points. How many 2-point baskets and 3-point baskets did Meagan make in the basketball game?**

Understand You know that she made both 2-point and 3-point baskets. You also know she scored a total of 9 points. You need to find how many of each she made.

Plan Make a guess until you find an answer that makes sense for the problem.

Solve

Number of 2-point baskets	Number of 3-point baskets	Total Number of Points
1	2	$1(2) + 2(3) = 8$
2	2	$2(2) + 2(3) = 10$
2	1	$2(2) + 1(3) = 7$
3	1	$3(2) + 1(3) = 9$

Check Three 2-point baskets result in 6 points. One 3-point basket results in 3 points. Since $6 + 3$ is 9, the answer is correct.

Exercise

VIDEO GAMES Juan has 16 video games. The types of video games he has are sports games, treasure hunts, and puzzles. He has 4 more sports games than treasure hunts. He has 3 fewer puzzles than treasure hunts. Use guess and check to determine how many of each type of video game Juan has.

Copyright © Glencoe/McGraw-Hill, a division of The McGraw-Hill Companies, Inc.

NAME ______________________ DATE ____________ PERIOD _____

1-7 Practice

Problem-Solving Investigation: Guess and Check

Mixed Problem Solving

Use the guess and check strategy to solve Exercises 1 and 2.

1. **MOVIES** Tickets for the movies are $7 for adults and $4 for children. Fourteen people paid a total of $68 for tickets. How many were adults and how many were children?

2. **AGES** Mei's mother is 4 times as old as Mei. Mei's grandmother is twice as old as Mei's mother. The sum of the three ages is 117. How old is Mei, her mother, and her grandmother?

Use any strategy to solve Exercises 3–6. Some strategies are shown below.

Problem-Solving Strategies
• Guess and check.
• Find a pattern.

3. **SWIMMING** Brian is preparing for a swim meet. The table shows the number of laps he swam in the first four days of practice. If the pattern continues, how many laps will Brian swim on Friday?

Day	Mon.	Tues.	Wed.	Thurs.	Fri.
Laps	1	3	7	15	?

4. **ORDER OF OPERATIONS** Use the symbols +, −, ×, and ÷ to make the following math sentence true. Write each symbol only once.

$8 \_\_ 2 \_\_ 1 \_\_ 3 \_\_ 4 = 5$

5. **PATTERNS** Draw the next figure in the pattern.

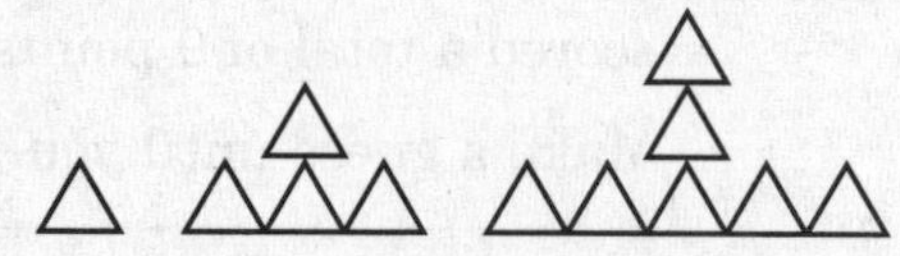

6. **MONEY** Jason has $1.56 in change in his pocket. If there is a total of 19 coins, how many quarters, dimes, nickels, and pennies does he have?

Copyright © Glencoe/McGraw-Hill, a division of The McGraw-Hill Companies, Inc.

NAME ______________ DATE ________ PERIOD ____

1-8 Study Guide and Intervention

Algebra: Equations

An **equation** is a sentence that contains an **equals sign**, =. Some equations contain variables. When you replace a variable with a value that results in a true sentence, you **solve** the equation. The value for the variable is the **solution** of the equation.

Example 1 **Solve $m + 12 = 15$ mentally.**

$m + 12 = 15$ Think: What number plus 12 equals 15?
$3 + 12 = 15$ You know that 12 + 3 = 15.
$m = 3$

The solution is 3.

Example 2 **Solve $14 - p = 6$ using guess and check.**

Guess the value of p, then check it out.

Try 7.	Try 6.	Try 8.
$14 - p \stackrel{?}{=} 6$	$14 - p \stackrel{?}{=} 6$	$14 - p \stackrel{?}{=} 6$
$14 - 7 \neq 6$	$14 - 6 \neq 6$	$14 - 8 = 6$
no	no	yes

The solution is 8 because replacing p with 8 results in a true sentence.

Exercises

Identify the solution of each equation from the list given.

1. $k - 4 = 13$; 16, 17, 18

2. $31 + x = 42$; 9, 10, 11

3. $45 = 24 + k$; 21, 22, 23

4. $m - 12 = 15$; 27, 28, 29

5. $88 = 41 + s$; 46, 47, 48

6. $34 - b = 17$; 16, 17, 18

7. $69 - j = 44$; 25, 26, 27

8. $h + 19 = 56$; 36, 37, 38

Solve each equation mentally.

9. $j + 3 = 9$

10. $m - 5 = 11$

11. $23 + x = 29$

12. $31 - h = 24$

13. $18 = 5 + d$

14. $35 - a = 25$

15. $y - 26 = 3$

16. $14 + n = 19$

17. $100 = 75 + w$

Lesson 1–8

Copyright © Glencoe/McGraw-Hill, a division of The McGraw-Hill Companies, Inc.

NAME ______________________ DATE __________ PERIOD _____

1-8 Practice

Algebra: Equations

Identify the solution of each equation from the list given.

1. $h + 9 = 21$; 10, 11, 12

2. $45 - k = 27$; 17, 18, 19

3. $34 + p = 52$; 18, 19, 20

4. $t \div 6 = 9$; 52, 53, 54

5. $43 = 52 - s$; 8, 9, 10

6. $56 = 7q$; 7, 8, 9

7. $28 = r - 12$; 40, 41, 42

8. $30 \div w = 5$; 4, 5, 6

9. $y - 13 = 24$; 37, 38, 39

Solve each equation mentally.

10. $a + 6 = 11$

11. $k - 12 = 4$

12. $24 = 34 - j$

13. $9b = 36$

14. $f \div 7 = 8$

15. $7 + n = 18$

16. $45 \div m = 5$

17. $80 = 10d$

18. $25 - c = 15$

19. $17 = 9 + e$

20. $g \div 4 = 12$

21. $26 \div k = 2$

22. ANIMALS A whiptail lizard has a tail that is twice as long as its body. The equation $2b = 6$ describes the length of a certain whiptail lizard's tail in inches. If b is the length of the whiptail lizard's body, what is the length of this whiptail lizard's body? What is the total length of the lizard?

23. SPORTS CAMP There are 475 campers returning to sports camp this year. Last year, 525 campers attended sports camp. The equation $475 = 525 - c$ shows the decrease in the number of campers returning to camp from one year to the next. Find the number of campers who did not return to camp this year.

Copyright © Glencoe/McGraw-Hill, a division of The McGraw-Hill Companies, Inc.

NAME ______________________ DATE ____________ PERIOD _____

1-9 Study Guide and Intervention

Algebra: Area Formulas

The **area** of a figure is the number of square units needed to cover a surface. You can use a formula to find the area of a rectangle. The formula for finding the area of a rectangle is $A = \ell \times w$. In this formula, A represents area, ℓ represents the length of the rectangle, and w represents the width of the rectangle.

Example 1 **Find the area of a rectangle with length 8 feet and width 7 feet.**

$A = \ell \times w$ Area of a rectangle
$A = 8 \times 7$ Replace ℓ with 8 and w with 7.
$A = 56$
The area is 56 square feet.

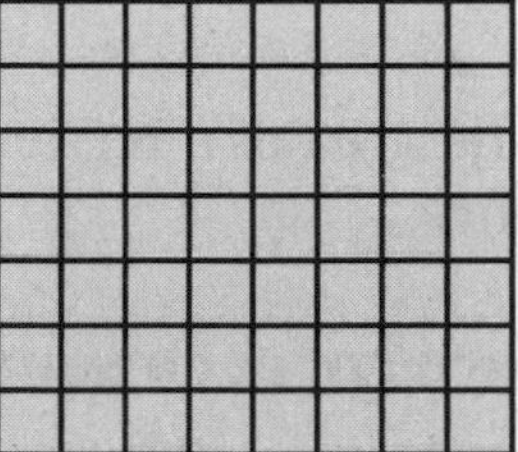

Example 2 **Find the area of a square with side length 5 inches.**

$A = s^2$ Area of a square
$A = 5^2$ Replace s with 5.
$A = 25$
The area is 25 square inches.

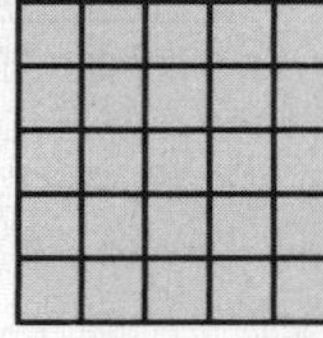

Exercises

Find the area of each figure.

1. 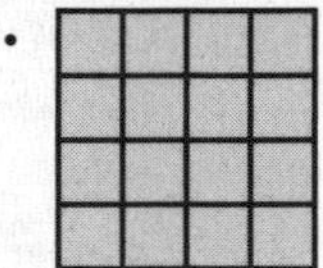

2. 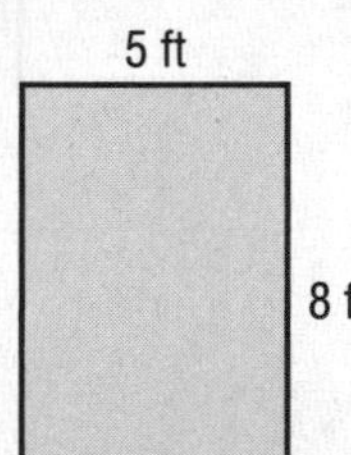

3. 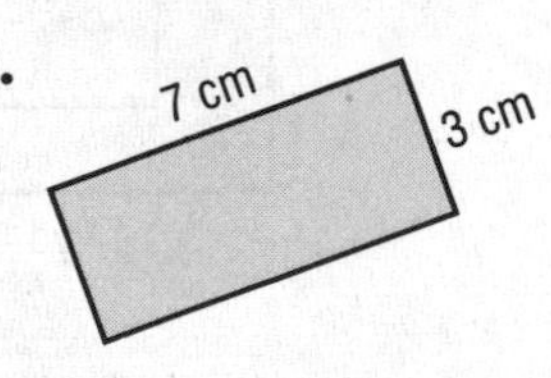

4. 

5. What is the area of a rectangle with a length of 10 meters and a width of 7 meters?

6. What is the area of a square with a side length of 15 inches?

Lesson 1–9

Copyright © Glencoe/McGraw-Hill, a division of The McGraw-Hill Companies, Inc.

NAME ______________________________ DATE ____________ PERIOD _____

1-9 Practice

Algebra: Area Formulas

Find the area of each rectangle.

1.

7 m

9 m

2.

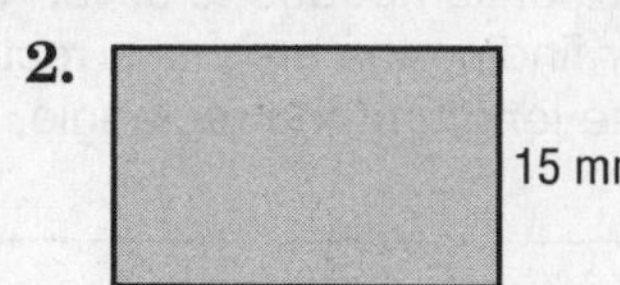

3.

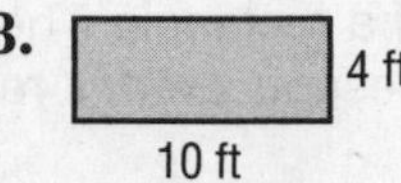

4. Find the area of a rectangle with a length of 35 inches and a width of 21 inches.

Find the area of each square.

5.

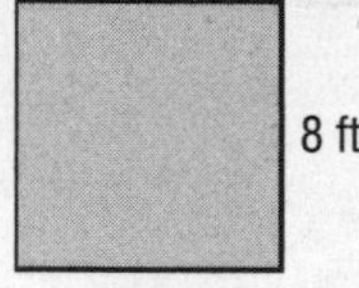

6.

7.

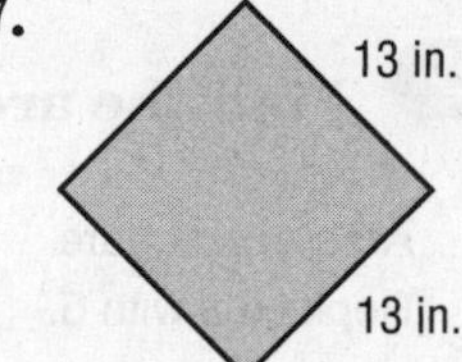

8. What is the area of a square with a side length of 21 yards?

Find the area of each shaded region.

10.

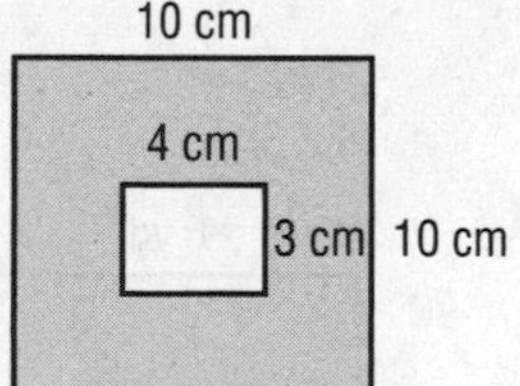

11.

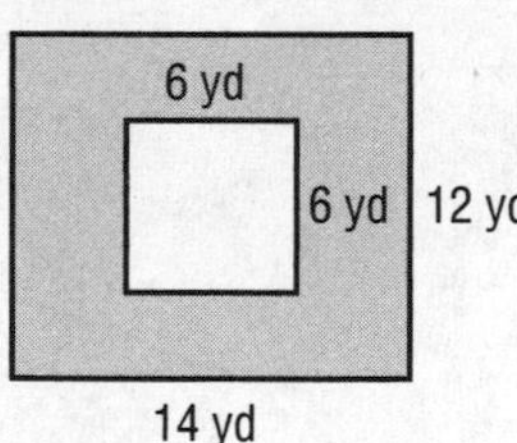

12.

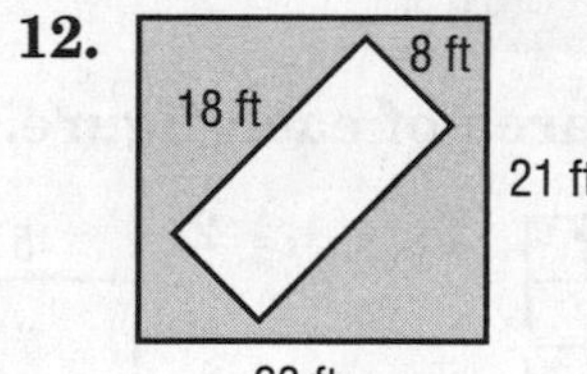

13. REMODELING The Crofts are covering the floor in their living room and in their bedroom with carpeting. The living room is 16 feet long and 12 feet wide. The bedroom is a square with 10 feet on each side. How many square feet of carpeting should the Crofts buy?

14. GARDENING The diagram shows a park's lawn with a sandy playground in the corner. If a bag of fertilizer feeds 5,000 square feet of lawn, how many bags of fertilizer are needed to feed the lawn area of the park?

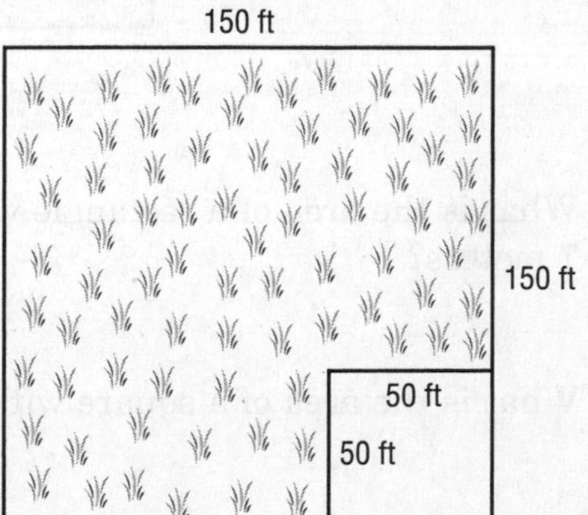

Copyright © Glencoe/McGraw-Hill, a division of The McGraw-Hill Companies, Inc.

NAME ______________________ DATE ____________ PERIOD _____

2-1 Study Guide and Intervention

Problem-Solving Investigation: Make a Table

When solving problems, one strategy that is helpful is to *make a table*. A table often makes it easy to clarify information in the problem. One type of table that is helpful to use is a *frequency table*, which shows the number of times each item or number appears.

You can use the *make a table* strategy, along with the following four-step problem-solving plan to solve a problem.

1 Understand – Read and get a general understanding of the problem.

2 Plan – Make a plan to solve the problem and estimate the solution.

3 Solve – Use your plan to solve the problem.

4 Check – Check the reasonableness of your solution.

Example 1

MOVIES Carlos took a survey of the students in his class to find out what type of movie they preferred. Using C for comedy, A for action, D for drama, and M for animated, the results are shown below. How many more students like comedies than action movies?

C A M M A C D C D C M A M M A C C D A C

Understand You need to find the number of students that chose comedies and the number of students that chose action. Then find the difference.

Plan Make a frequency table of the data.

Solve Draw and complete a frequency table.

7 people chose comedies and 5 people chose action. So, 7 − 5 or 2 more students chose comedy than action.

Check Go back to the list to verify there are 7 C's for comedy and 5 A's for action.

Favorite Type of Movie		
Movie Type	**Tally**	**Frequency**
comedy	𝍸 II	7
action	𝍸	5
drama	III	3
animated	𝍸	5

Exercise

GRADES The list below shows the quarterly grades for Mr. Vaquera's math class. Make a frequency table of the data. How many more students received a B than a D?

B C A A B D C B A C B B
B D A C B B C A A B A B

Lesson 2–1

Copyright © Glencoe/McGraw-Hill, a division of The McGraw-Hill Companies, Inc.

2-1 Practice

Problem-Solving Investigation: Make a Table

Mixed Problem Solving

Use the make a table strategy to solve Exercise 1.

1. **BASKETBALL** The winning scores for teams in the National Wheelchair Basketball Association junior division for a recent season are shown. Make a frequency table of the data. How many winning scores were between 21 and 25?

NWBA Jr. Div. Winning Scores					
25	26	34	16	33	18
34	26	24	33	12	23

Use any strategy to solve Exercises 2–5. Some strategies are shown below.

Problem-Solving Strategies
• Guess and check.
• Make a table.

2. **MONEY** Emelio has 9 coins that total $2.21. He does not have a dollar coin. What are the coins?

3. **SCIENCE** A biologist counted the birds she tagged and released each day for 20 days. Her counts were: 13, 14, 9, 16, 21, 8, 28, 25, 9, 13, 23, 16, 14, 9, 21, 25, 8, 10, 21, and 29. On how many days did she count between 6 and 10 birds or between 26 and 30 birds?

4. **TRAFFIC** The table shows the types of vehicles seen passing a street corner. Make a frequency table of the data. How many fewer motorcycles than cars were seen?

Types of Vehicles							
C	M	M	B	T	T	C	T
B	R	T	C	R	C	R	C
M	C	C	M	C	R	C	T

C = car B = bicycle T = truck

M = motorcycle R = recreational vehicle

5. **MONEY** Tonisha has $0 in her savings account. She deposits $40 every two weeks and withdraws $25 every four weeks. What will be her balance at the end of 24 weeks?

Copyright © Glencoe/McGraw-Hill, a division of The McGraw-Hill Companies, Inc.

NAME ______________________ DATE ____________ PERIOD _____

2-2 Study Guide and Intervention

Bar Graphs and Line Graphs

A **graph** is a visual way to display data. A **bar graph** is used to compare data. A **line graph** is used to show how data changes over a period of time.

Example 1 **Make a bar graph of the data. Compare the number of students in jazz class with the number in ballet class.**

Dance Classes	
Style	**Students**
Ballet	11
Tap	4
Jazz	5
Modern	10

Step 1 Decide on the scale and interval.

Step 2 Label the horizontal and vertical axes.

Step 3 Draw bars for each style.

Step 4 Label the graph with a title.

About twice as many students take ballet as take jazz.

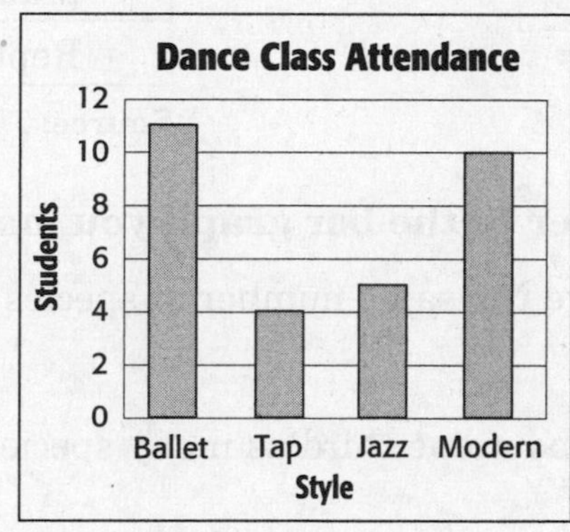

Example 2 **Make a line graph of the data. Then describe the change in Gwen's allowance from 2003 to 2008.**

Gwen's Allowance						
Year	2003	2004	2005	2006	2007	2008
Amount ($)	10	15	15	18	20	25

Gwen's Allowance

Step 1 Decide on the scale and interval.

Step 2 Label the horizontal and vertical axes.

Step 3 Draw and connect the points for each year.

Step 4 Label the graph with a title.

Gwen's allowance did not change from 2004 to 2005 and then increased from 2005 to 2008.

Exercises

Make the graph listed for each set of data.

1. bar graph

Riding the Bus	
Student	**Time (min)**
Paulina	10
Omar	40
Ulari	20
Jacob	15
Amita	35

2. line graph

Getting Ready for School	
Day	**Time (min)**
Monday	34
Tuesday	30
Wednesday	37
Thursday	20
Friday	25

Copyright © Glencoe/McGraw-Hill, a division of The McGraw-Hill Companies, Inc.

NAME ______________________ DATE ____________ PERIOD _____

2-2 Practice

Bar Graphs and Line Graphs

1. **ANIMALS** Make a bar graph of the data.

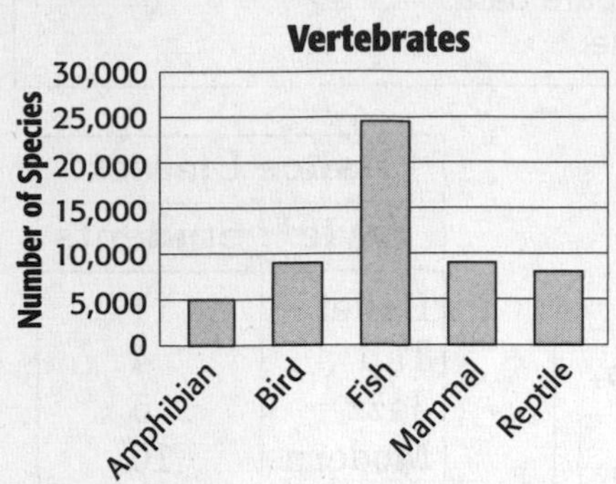

Vertebrates	
Class	**Number of Species**
Amphibians	5,000
Birds	9,000
Fish	24,500
Mammals	9,000
Reptiles	8,000

Source: *The World Almanac for Kids*

For Exercises 2 and 3, refer to the bar graph you made in Exercise 1.

2. Which animal classes have the same number of species?

3. Which animal class has about one third as many species as the fish class?

4. **POPULATION** Make a line graph of the data.

Population of the District of Columbia	
Year	**Population (thousands)**
1960	764
1970	757
1980	638
1990	607
2000	572

Source: U.S. Census Bureau

For Exercises 5 and 6, refer to the line graph you made in Exercise 4.

5. Describe the change in the District of Columbia population from 1970 to 2000.

6. What year showed the greatest change in population from the previous year?

BOOKS For Exercises 7 and 8, refer to the table.

7. Choose an appropriate scale and interval for the data set.

8. Would this data set be best represented by a bar graph or a line graph? Explain your reasoning.

Book Sales			
Week	**Sales ($)**	**Week**	**Sales ($)**
1	110	5	40
2	118	6	103
3	89	7	30
4	74	8	58

Copyright © Glencoe/McGraw-Hill, a division of The McGraw-Hill Companies, Inc.

NAME ______________________ DATE __________ PERIOD _____

2-3 Study Guide and Intervention

Interpret Line Graphs

Because they show trends over time, **line graphs** are often used to predict future events.

Example 1 **The graph shows the time Ruben spends each day practicing piano scales. Predict how much time he will spend practicing his scales on Friday.**

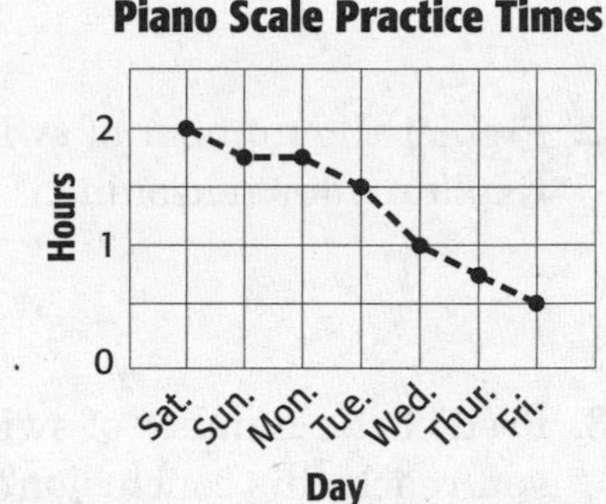

Continue the graph with a dotted line in the same direction until you reach a vertical position for Friday. By extending the graph, you see that Ruben will probably spend half an hour practicing piano scales on Friday.

Exercises

MONEY **Use the graph that shows the price of a ticket to a local high school football game over the last few years.**

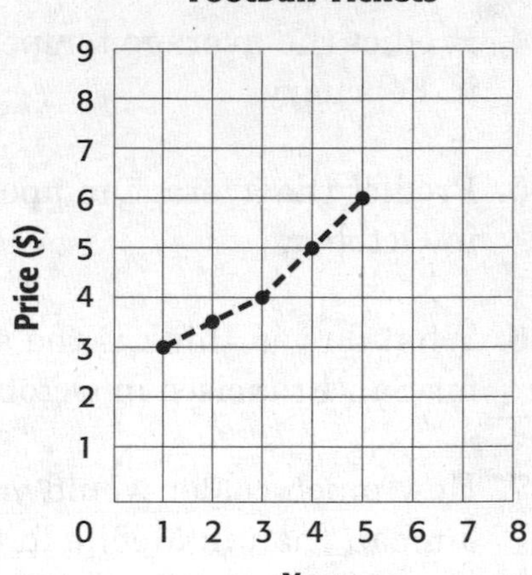

1. Has the price been increasing or decreasing? Explain.

2. Predict the price of a ticket in year 6 if the trend continues.

3. In what year do you think the price will reach \$9.00 if the trend continues?

BANKS **Use the graph that shows the interest rate for a savings account over the last few years.**

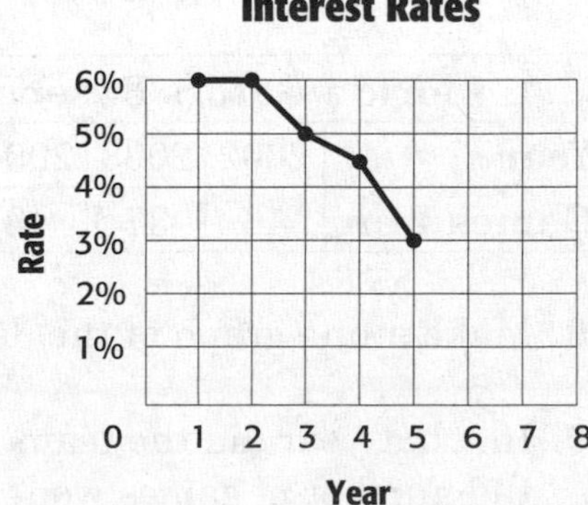

4. What does the graph tell you about interest rates?

5. If the trend continues, when will the interest rate reach 1 percent?

Copyright © Glencoe/McGraw-Hill, a division of The McGraw-Hill Companies, Inc.

Lesson 2-3

NAME ______________________ DATE __________ PERIOD ______

2-3 Practice

Interpret Line Graphs

SPORTS For Exercises 1–3, use the graph at the right.

Swimsuit Sales

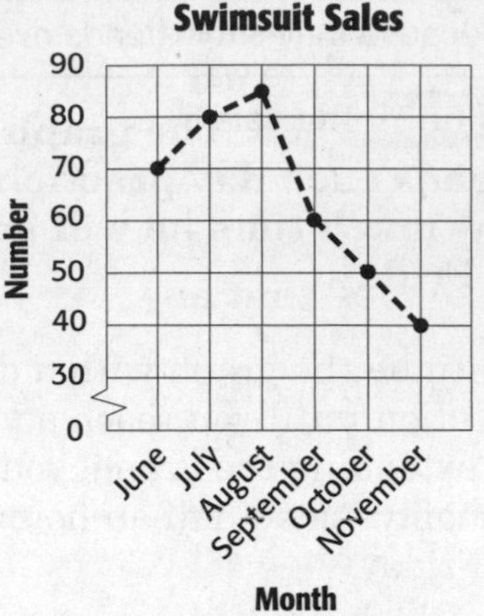

1. Describe the change in the number of swimsuits sold.

2. Predict the number of swimsuits sold in December. Explain your reasoning.

3. Predict the number of swimsuits sold in May. How did you reach this conclusion?

WEATHER For Exercises 4–7, use the graph at the right.

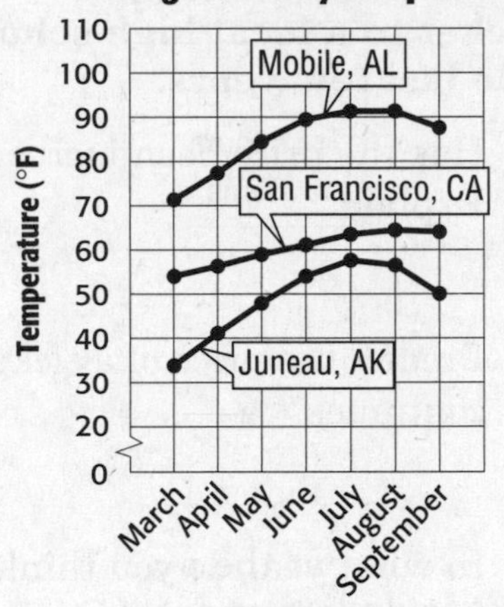

Source: *The World Almanac*

4. Predict the average temperature for Juneau in February.

5. Predict the average temperature for Mobile in October.

6. What do you think is the average temperature for San Francisco in October?

7. How much colder would you expect it to be in Juneau than in Mobile in October?

BASEBALL For Exercises 8–10, use the table that shows the number of games won by the Florida Gators men's baseball team from 2002 to 2007.

Florida Gators Baseball Statistics

Year	2002	2003	2004	2005	2006	2007
Games Won	46	37	43	48	28	29

8. Make a line graph of the data.

9. In what year did the team have the greatest increase in the number of games won?

10. Explain the disadvantages of using this line graph to make a prediction about the number of games that the team will win in 2009.

Copyright © Glencoe/McGraw-Hill, a division of The McGraw-Hill Companies, Inc.

2-4 Study Guide and Intervention

Stem-and-Leaf Plots

Sometimes it is hard to read data in a table. You can use a **stem-and-leaf plot** to display the data in a more readable way. In a stem-and-leaf plot, you order the data from least to greatest. Then you organize the data by place value.

Example 1 **Make a stem-and-leaf plot of the data in the table. Then write a few sentences that analyze the data.**

Money Earned Mowing Lawns ($)			
60	55	53	41
67	72	65	68
65	70	52	51

Step 1 Order the data from least to greatest.
41 51 52 53 55 60 65 65 67 68 70 72

Step 2 Draw a vertical line and write the tens digits from least to greatest to the left of the line.

Step 3 Write the ones digits to the right of the line with the corresponding stems.

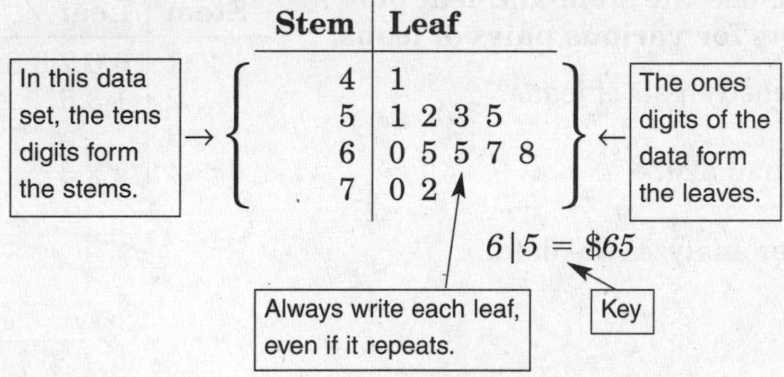

Step 4 Include a key that explains the stems and leaves.

By looking at the plot, it is easy to see that the least amount of money earned was $41 and the greatest amount was $72. You can also see that most of the data fall between $51 and $68.

Exercise

Make a stem-and-leaf plot for the set of data below. Write a few sentences that analyze the data.

Points scored: 34 44 51 48 55 41 47 22 55

Copyright © Glencoe/McGraw-Hill, a division of The McGraw-Hill Companies, Inc.

NAME ______________________ DATE __________ PERIOD _____

2-4 Practice

Stem-and-Leaf Plots

Make a stem-and-leaf plot for each set of data.

1. Minutes on the bus to school: 10, 5, 21, 30, 7, 12, 15, 21, 8, 12, 12, 20, 31, 10, 23, 31

2. Employee's ages: 22, 52, 24, 19, 25, 36, 30, 32, 19, 26, 28, 33, 53, 24, 35, 26

SHOPPING **For Exercises 3–5, use the stem-and-leaf plot at the right that shows costs for various pairs of jeans.**

Stem	Leaf
1	6 6 7 8 8 9 9 9 9
2	1 3 5
3	
4	2 2 3

2 | 3 = $23

3. How much is the most expensive pair of jeans?

4. How many pairs cost less than $20?

5. Write a sentence or two that analyzes the data.

6. Construct a stem-and-leaf plot for the set of test scores 81, 55, 55, 62, 73, 49, 56, 91, 55, 64, 72, 62, 64, 53, 56, and 57. Then write sentences explaining how a teacher might use the plot.

7. Display the amounts $104, $120, $99, $153, $122, $116, $114, $139, $102, $95, $123, $116, $152, $104 and $115 in a stem-and-leaf plot. (*Hint*: Use the hundreds and tens digits to form the stems.)

Copyright © Glencoe/McGraw-Hill, a division of The McGraw-Hill Companies, Inc.

NAME ______________________________ DATE ______________ PERIOD ________

2-5 Study Guide and Intervention

Line Plots

A **line plot** is a diagram that shows the frequency of data on a number line. A line plot is created by drawing a number line and then placing an × above a data value each time that data occurs.

Example 1 **Make a line plot of the data in the table at the right.**

Time Spent Traveling to School (minutes)						
5	6	3	10	12	15	5
10	5	8	12	5	5	8

Draw a number line. The smallest value is 3 minutes and the largest value is 15 minutes. So, you can use a scale of 0 to 15.

Put an × above the number that represents the travel time of each student in the table. Be sure to include a title.

0 5 10 15

Example 2 **How many students spend 5 minutes traveling to school each day?**

Locate 5 on the number line and count the number of ×'s above it. There are 5 students that travel 5 minutes to school each day.

Exercises

AGES For Exercises 1–3, use the data below.

Ages of Lifeguards at Brookville Swim Club					
16	18	16	20	22	18
18	17	18	25	17	19

1. Make a line plot of the data.

2. How many of the lifeguards are 18 years old?

3. What is the age difference between the oldest and youngest lifeguard at Brookville Swim Club?

Lesson 2–5

Copyright © Glencoe/McGraw-Hill, a division of The McGraw-Hill Companies, Inc.

NAME ______________________ DATE ____________ PERIOD _____

2-5 Practice

Line Plots

PRESIDENTS For Exercises 1–4, use the line plot below. It shows the ages of the first ten Presidents of the United States when they first took office.

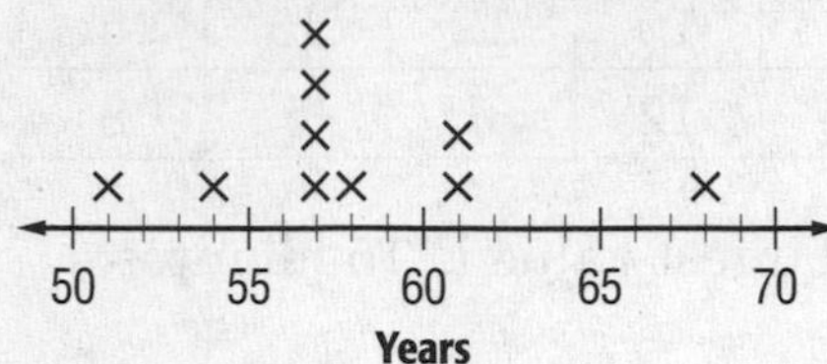

Source: *Time Almanac*

1. How many Presidents were 54 when they took office?

2. Which age was most common among the first ten Presidents when they took office?

3. How many Presidents were in their 60s when they first took office?

4. What is the difference between the age of the oldest and youngest President represented in the line plot?

5. **EXERCISE** Make a line plot for the set of data.

Miles Walked this Week			
16	21	11	24
8	14	16	11
21	10	8	14
11	24	12	18
18	27	11	14

BIRDS For Exercises 6 and 7, use the line plot below. It shows the number of mockingbirds each bird watcher saw on a bird walk.

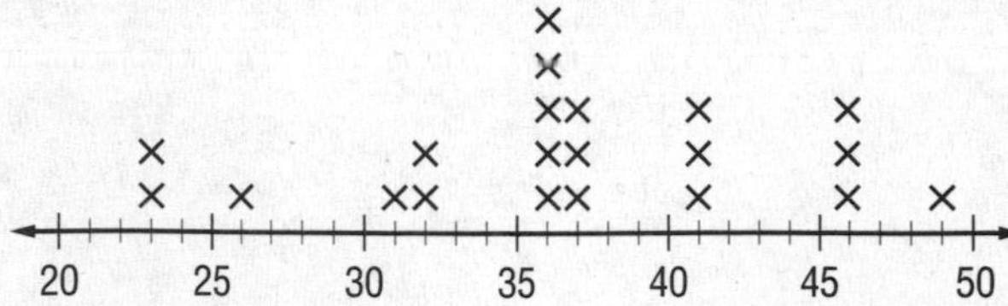

6. How many more bird watchers saw 36 mockingbirds than saw 46 mockingbirds?

7. How many bird watchers are represented in the line plot?

Copyright © Glencoe/McGraw-Hill, a division of The McGraw-Hill Companies, Inc.

NAME ______________________ DATE ____________ PERIOD ______

2-6 Study Guide and Intervention

Mean

The **mean** is the most common measure of central tendency. It is an average, so it describes all of the data in a data set.

Example 1 **The picture graph shows the number of members on four different swim teams. Find the mean number of members for the four different swim teams.**

Swim Team Members	
Amberly	
Carlton	
Hamilton	
Westhigh	

Simplify an expression.

$$\text{mean} = \frac{9 + 11 + 6 + 10}{4}$$

$$= \frac{36}{4} \text{ or } 9$$

A set of data may contain very high or very low values. These values are called **outliers**.

Example 2 **Find the mean for the snowfall data with and without the outlier. Then tell how the outlier affects the mean of the data.**

Month	Snowfall (in.)
Nov.	20
Dec.	19
Jan.	20
Feb.	17
Mar.	4

Compared to the other values, 4 inches is low. So, it is an outlier.

mean with outlier

$$\text{mean} = \frac{20 + 19 + 20 + 17 + 4}{5}$$

$$= \frac{80}{5} \text{ or } 16$$

mean without outlier

$$\text{mean} = \frac{20 + 19 + 20 + 17}{4}$$

$$= \frac{76}{4} \text{ or } 19$$

With the outlier, the mean is less than the values of most of the data. Without the outlier, the mean is close in value to the data.

Exercises

SHOPPING **For Exercises 1–3, use the bar graph at the right.**

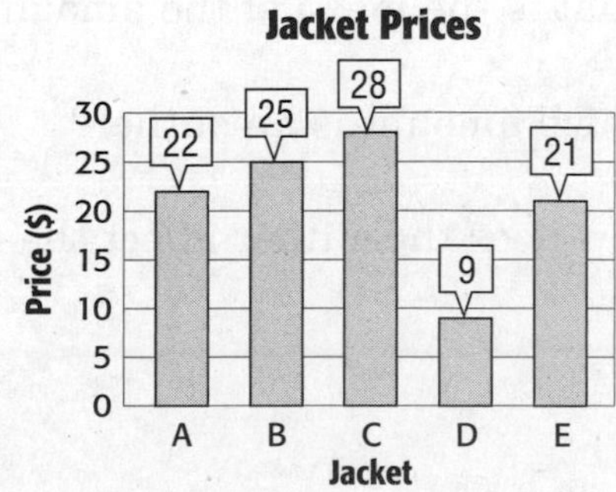

1. Find the mean of the data.

2. Which jacket price is an outlier?

3. Find the mean of the data if the outlier is not included.

4. How does the outlier affect the mean of the data?

Lesson 2–6

Copyright © Glencoe/McGraw-Hill, a division of The McGraw-Hill Companies, Inc.

NAME ______________________ DATE ____________ PERIOD _____

2-6 Practice

Mean

Find the mean of the data represented in each model.

1.

Number of Toys Collected	
Ling	
Kathy	
Lucita	
Terrell	

2.

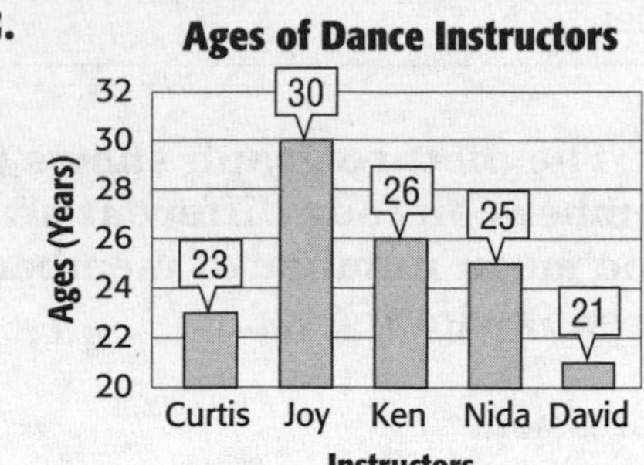

NATURE For Exercises 3–6, use the table that shows the heights of the tallest waterfalls along Oregon's Columbia River Gorge.

Falls	Height (ft)
Bridal Veil	153
Horsetail	176
Latourell	249
Metlako	150
Multnomah	620
Wahkeena	242

3. Find the mean of the data.

4. Identify the outlier.

5. Find the mean if Multnomah Falls is not included in the data set.

6. How does the outlier affect the mean of the data?

GARDENING For Exercises 7–9, use the following information.

Alan earned \$23, \$26, \$25, \$24, \$23, \$24, \$6, \$24, and \$23 gardening.

7. What is the mean of the amounts he earned?

8. Which amount is an outlier?

9. How does the outlier affect the mean of the data?

Find the mean for number of cans collected. Explain the method you used.

10. 57, 59, 60, 58, 58, 56

Copyright © Glencoe/McGraw-Hill, a division of The McGraw-Hill Companies, Inc.

NAME ______________________ DATE ____________ PERIOD _____

2-7 Study Guide and Intervention

Median, Mode, and Range

The **median** is the middle number of the data put in order, or the mean of the middle two numbers.
The **mode** is the number or numbers that occur most often.

Example 1 **The table shows the costs of seven different books. Find the mean, median, and mode of the data.**

Book Costs ($)			
22	13	11	16
14	13	16	

mean: $\frac{22 + 13 + 11 + 16 + 14 + 13 + 16}{7} = \frac{105}{7}$ or 15

To find the median, write the data in order from least to greatest.
median: 11, 13, 13, (14), 16, 16, 22

To find the mode, find the number or numbers that occur most often.
mode: 11, (13, 13), 14, (16, 16), 22

The mean is $15. The median is $14. There are two modes, $13 and $16.

Whereas the measures of central tendency describe the average of a set of data, the **range** of a set of data describes how the data vary.

Example 2 **Find the range of the data in the stem-and-leaf plot. Then write a sentence describing how the data vary.**

Stem	Leaf
3	2
4	0
5	0 5
6	0 3

3|2 = 32°

The greatest value is 63. The least value is 32. So, the range is 63° − 32° or 31°. The range is large. It tells us that the data vary greatly in value.

Exercises

Find the mean, median, mode, and range of each set of data.

1. hours worked: 14, 13, 14, 16, 8

2. points scored by football team: 29, 31, 14, 21, 31, 22, 20

3.

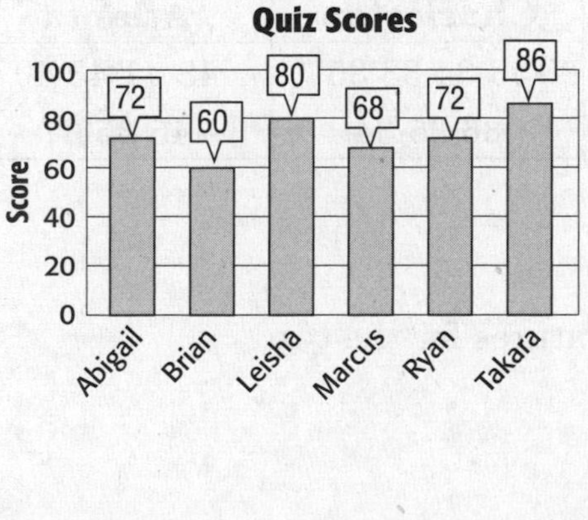

4.

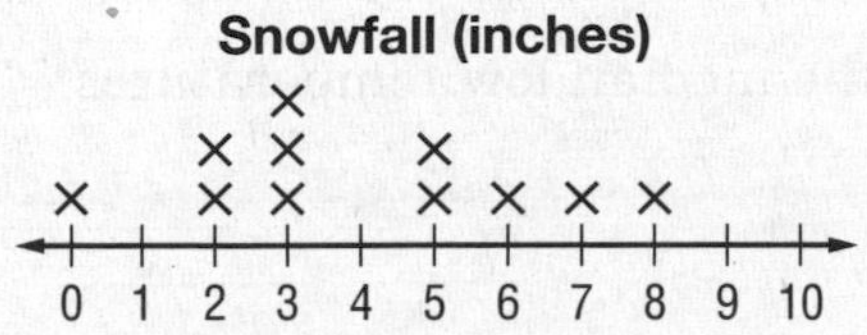

Copyright © Glencoe/McGraw-Hill, a division of The McGraw-Hill Companies, Inc.

Lesson 2–7

NAME ______________________________ DATE ____________ PERIOD _____

2-7 Practice

Median, Mode, and Range

Find the median, mode, and range for each set of data.

1. minutes spent practicing violin: 25, 15, 30, 25, 20, 15, 24

2. snow in inches: 40, 28, 24, 37, 43, 26, 30, 36

Find the mean, median, mode, and range of the data represented in each statistical graph.

3.

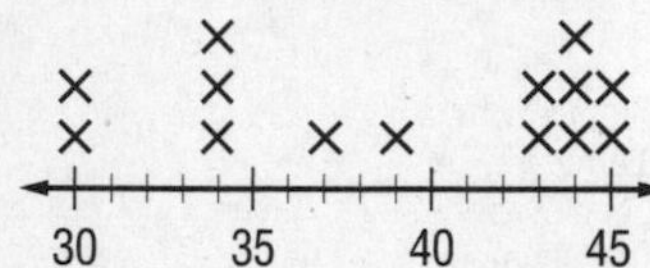

4.

Stem	Leaf
4	1 2 4 4
5	2 4
6	1 3 4 7 7 7 7 8 8
7	2 2 3
8	0 1 2 4 5 6

5|4 = $54

5.

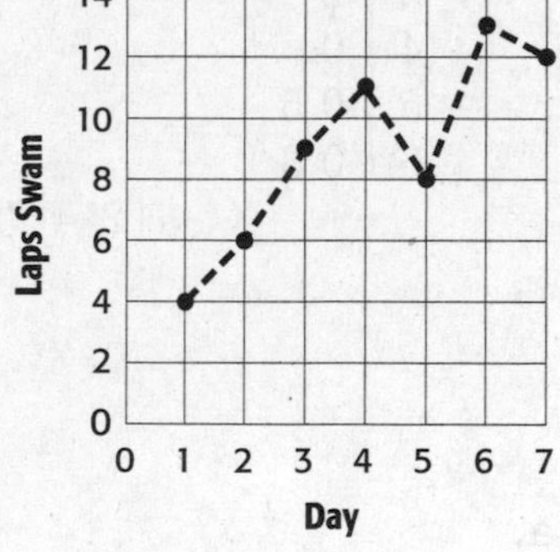

6.

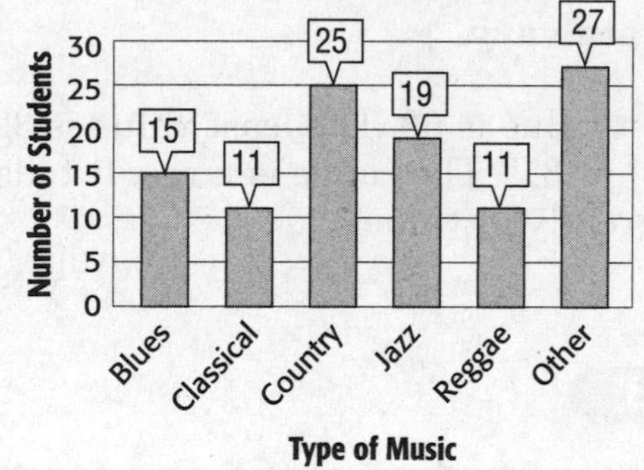

WEATHER For Exercises 7–9, refer to the table at the right.

7. Compare the median low temperatures.

Daily Low Temperatures (°F)	
Charleston	**Atlanta**
33 34 33 35	48 41 43 40
36 35 34	45 35 37

8. Find the range for each data set.

9. Write a statement that compares the daily low temperatures for the two cities.

Copyright © Glencoe/McGraw-Hill, a division of The McGraw-Hill Companies, Inc.

NAME ______________________ DATE ____________ PERIOD _____

2-8 Study Guide and Intervention

Selecting an Appropriate Display

Data can be displayed in many different ways, including the following:

- A **bar graph** shows the number of items in a specific category.
- A **line graph** shows change over a period of time.
- A **line plot** shows how many times each number occurs in the data.
- A **stem-and-leaf plot** lists all individual numerical data in a condensed form.

Example 1 **Which display allows you to see how art show ticket prices have changed since 2004.**

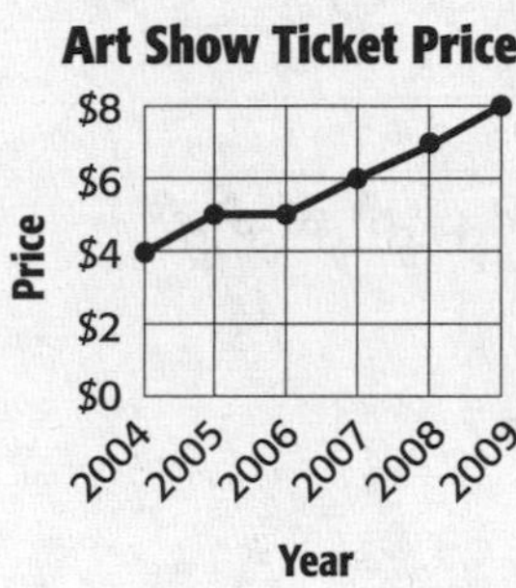

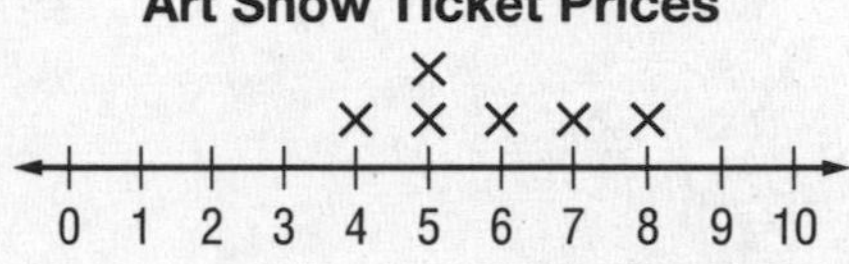

The line graph allows you to see how the art show ticket prices have increased since 2004.

Example 2 **What type of display would you use to show the results of a survey of students' favorite brand of tennis shoes.**

Since the data would list the number of students that chose each brand, or category, the data would best be displayed in a bar graph.

Exercises

1. **GRADES** Which display makes it easier to see how many students had test scores in the 80s?

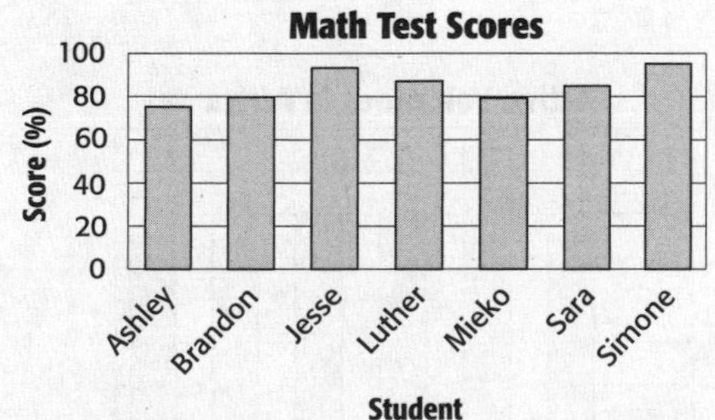

Math Test Scores

Stem	Leaf
7	5 9
8	0 5 8
9	2 3

8|0 = 80%

2. **VOLLEYBALL** What type of display would you use to show the number of wins the school volleyball team had from 2000 to 2005?

Copyright © Glencoe/McGraw-Hill, a division of The McGraw-Hill Companies, Inc.

NAME ______________________ DATE __________ PERIOD ______

2-8 Practice

Selecting an Appropriate Display

1. **FOOD** **Which display makes it easier to see the median cost of providing food stamps from 1998 to 2003?**

Stem	Leaf
1	7 8 8 9
2	1 4

1|7 = 17 thousand million dollars

Source: *The World Almanac*

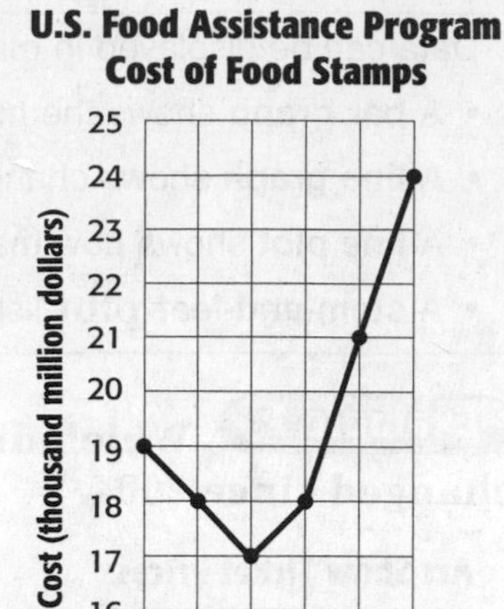

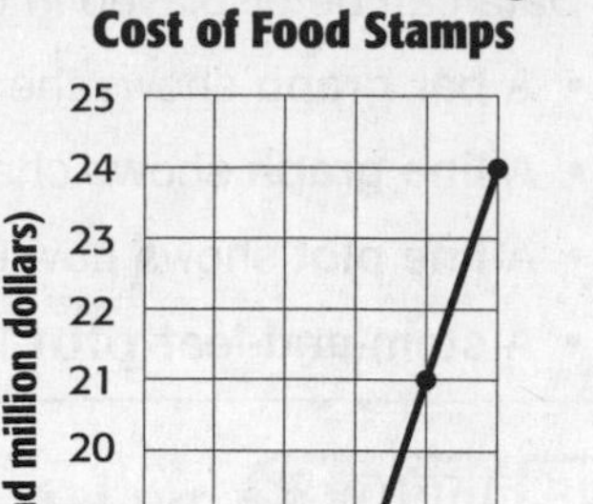

Select an appropriate type of display for data gathered about each situation. Sample answers are given.

2. the heights of buildings in town
3. the number of cars a dealer sold each month over the past year
4. the number of scores made by each team member in a basketball season
5. **OLYMPICS** Select an appropriate type of display for the data. Then make a display.

Olympic Hammer Throw Winners			
Year	**Distance (m)**	**Year**	**Distance (m)**
1968	73	1988	85
1972	76	1992	83
1976	78	1996	81
1980	82	2000	80
1984	78	2004	83

6. **GEOGRAPHY** Display the data in the bar graph using another type of display. Compare the displays.

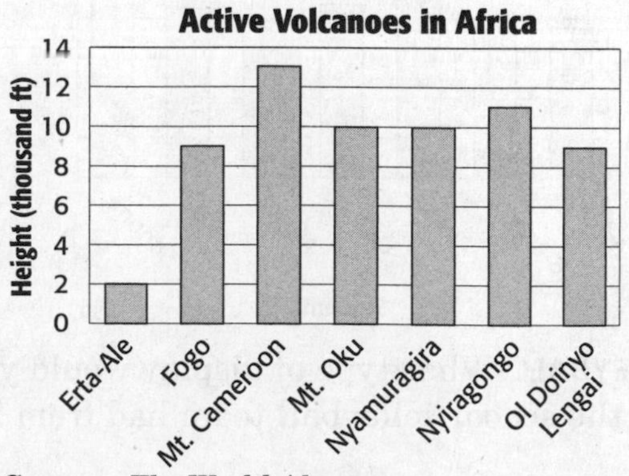

Source: *The World Almanac*

Copyright © Glencoe/McGraw-Hill, a division of The McGraw-Hill Companies, Inc.

NAME ______________________________ DATE ______________ PERIOD ______

2-9 Study Guide and Intervention

Integers and Graphing

Negative numbers represent data that are less than 0. A negative number is written with a − sign. **Positive numbers** represent data that are greater than 0. Positive numbers are written with a + sign or no sign at all.

Opposites are numbers that are the same distance from zero on a number line, but in opposite directions. The set of positive whole numbers, their opposites, and zero are called **integers**.

Example 1 **Write an integer to show 3 degrees below zero. Then graph the integer on a number line.**

Numbers *below zero* are negative numbers. The integer is −3.

Draw a number line. Then draw a dot at the location that represents −3.

−6 −5 −4 −3 −2 −1 0 1 2 3 4 5 6

Example 2 **Make a line plot of the data represented in the table.**

Draw a number line. Put an × above the number that represents each score in the table.

Rachel's Summer Golf Scores			
0	+3	−4	−2
+1	+3	−4	0
+1	−5	−2	+1

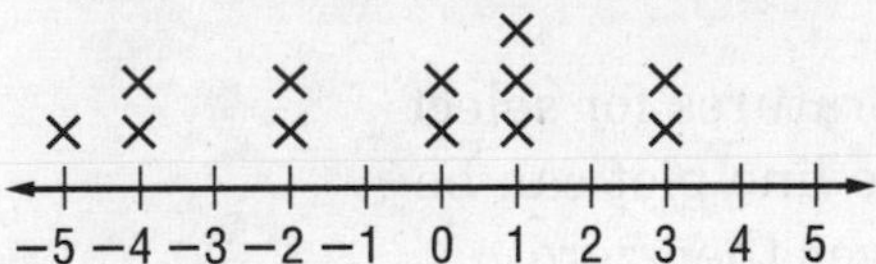

Exercises

Write an integer to represent each piece of data. Then graph the integer on the number line.

1. 4 degrees below zero

−6 −5 −4 −3 −2 −1 0 1 2 3 4 5 6

2. a gain of 2 points

3. BOOKS The table shows the change in the ranking from the previous week of the top ten best-selling novels. Make a line plot of the data.

Novel	A	B	C	D	E	F	G	H	I	J
Change in Ranking	+3	−2	0	+1	−2	0	+2	−4	+1	−2

Copyright © Glencoe/McGraw-Hill, a division of The McGraw-Hill Companies, Inc.

Lesson 2-9

NAME ______________________________ DATE ____________ PERIOD _____

2-9 Practice

Integers and Graphing

Write an integer to represent each situation.

1. Bill drove 25 miles toward Tampa.

2. Susan lost $4.

3. Joe walked down 6 flights of stairs.

4. The baby gained 8 pounds.

Draw a number line from −10 to 10. Then graph each integer on the number line.

5. 2 **6.** 6 **7.** 10 **8.** 8

9. −7 **10.** −4 **11.** −9 **12.** −3

Write the opposite of each integer.

13. +8 **14.** −5 **15.** −2 **16.** +9

17. −11 **18.** +21 **19.** +10 **20.** −7

21. **SCIENCE** The average daytime surface temperature on the Moon is 260°F. Represent this temperature as an integer.

22. **GEOGRAPHY** The Salton Sea is a lake at 227 feet below sea level. Represent this altitude as an integer.

23. **WEATHER** The table below shows the extreme low temperatures for select cities. Make a line plot of the data. Then explain how the line plot can be used to determine whether more cities had extremes lower then zero degrees or greater than zero degrees.

Extreme Low Temperatures by City

City	Temp. °F	City	Temp. °F
Mobile, AL	3	Boston, MA	−12
Wilmington, DE	−14	Jackson, MS	2
Jacksonville, FL	7	Raleigh, NC	−9
Savannah, GA	3	Portland, OR	−3
New Orleans, LA	11	Philadelphia, PA	−7
Baltimore, MD	−7	Charleston, SC	6

Source: *The World Almanac*

Copyright © Glencoe/McGraw-Hill, a division of The McGraw-Hill Companies, Inc.

NAME ______________________ DATE ____________ PERIOD _____

3-1 Study Guide and Intervention

Representing Decimals

Lesson 3–1

Decimals can be written in standard form and expanded form.

Standard form is the usual way to write a decimal, such as 3.52. **Expanded form** is a sum of the products of each digit and its place, such as (3 × 1) + (5 × 0.1) + (2 × 0.01).

Example 1 **Write 128.0732 in word form.**

Place-Value Chart							
thousands	hundreds	tens	ones	tenths	hundredths	thousandths	ten-thousandths
0	1	2	8 .	0	7	3	2

In words, 128.0732 is *one hundred twenty-eight and seven hundred thirty-two ten-thousandths.*

Example 2 **Write *ninety-nine and two hundred seven thousandths* in standard form and expanded form.**

Place-Value Chart							
thousands	hundreds	tens	ones	tenths	hundredths	thousandths	ten-thousandths
0	0	9	9 .	2	0	7	0

Standard form: 99.207
Expanded form: (9 × 10) + (9 × 1) + (2 × 0.1) + (0 × 0.01) + (7 × 0.001)

Exercises

Write each decimal in word form.

1. 2.3

2. 0.68

3. 32.501

4. 0.0036

Write each decimal in standard form and in expanded form.

5. twenty and two hundredths

6. seven and five tenths

7. three hundred four ten-thousandths

8. eleven thousandths

Copyright © Glencoe/McGraw-Hill, a division of The McGraw-Hill Companies, Inc.

NAME ______________________ DATE ____________ PERIOD _____

3-1 Practice

Representing Decimals

Write each decimal in word form.

1. 0.5

2. 0.1

3. 2.49

4. 8.07

5. 0.345

6. 30.089

7. 6.0735

8. 0.0042

9. 16.375

Write each decimal in standard form and in expanded form.

10. one tenth

11. thirteen and four tenths

12. sixty-two and thirty-five hundredths

13. seven hundred twelve ten-thousandths

14. How is 611.0079 written in word form?

15. Write $(2 \times 0.1) + (8 \times 0.01)$ in word form.

16. Write $(5 \times 0.001) + (6 \times 0.0001)$ in standard form.

17. **HIKING** Pinnacles National Monument in California has 71.2 miles of hiking trails. Write this number in two other forms.

18. **ANALYZE TABLES** In the table at the right, which numbers have their last digit in the thousandths place? Explain your reasoning. Write each of these numbers in expanded form.

World Records For Smallest Animal

Animal	Length (cm)
dog	7.112
hamster	4.445
newt	2.54
spider	0.0432
starfish	0.889
toad	2.3876

Source: *Guinness World Records*

Copyright © Glencoe/McGraw-Hill, a division of The McGraw-Hill Companies, Inc.

3-2 Study Guide and Intervention

Comparing and Ordering Decimals

Example 1 **Use > or < to compare 68.563 and 68.5603.**

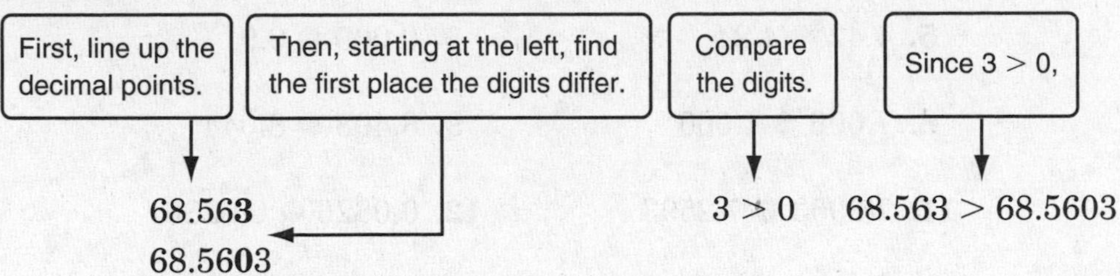

So, 68.563 is greater than 68.5603.

Example 2 **Order 4.073, 4.73, 4.0073, and 4 from least to greatest.**

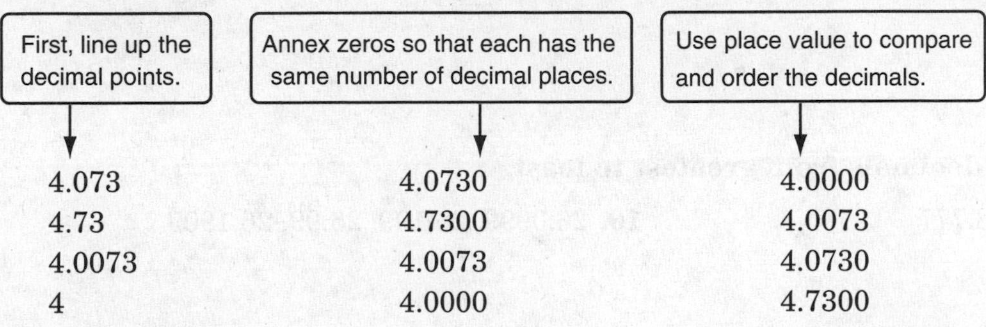

The order from least to greatest is 4, 4.0073, 4.073, and 4.73.

Exercises

Use >, <, or = to compare each pair of decimals.

1. 4.08 ● 4.080 **2.** 0.001 ● 0.01 **3.** 23.659 ● 22.659

4. 50.031 ● 50.030 **5.** 7 ● 7.0001 **6.** 18.01 ● 18.010

Order each set of decimals from least to greatest.

7. 0.006, 0.6, 0.060, 6 **8.** 456.73, 465.32, 456.37, 456.23

Order each set of decimals from greatest to least.

9. 3.01, 3.009, 3.09, 3.0001 **10.** 45.303, 45.333, 45.03, 45.0003, 45.003

Copyright © Glencoe/McGraw-Hill, a division of The McGraw-Hill Companies, Inc.

NAME ______________________________ DATE ____________ PERIOD _____

3-2 Practice

Comparing and Ordering Decimals

Use >, <, or = to compare each pair of decimals.

1. 8.8 ● 8.80 **2.** 0.3 ● 3.0 **3.** 0.06 ● 0.6

4. 5.10 ● 5.01 **5.** 4.42 ● 4.24 **6.** 0.009 ● 0.9

7. 0.305 ● 0.315 **8.** 7.006 ● 7.060 **9.** 8.408 ● 8.044

10. 91.77 ● 91.770 **11.** 7.2953 ● 7.2593 **12.** 0.0826 ● 0.0286

Order each set of decimals from least to greatest.

13. 33.6, 34.01, 33.44, 34 **14.** 78.203, 78.34, 78.023, 78.23

Order each set of decimals from greatest to least.

15. 8.7, 8.77, 8.07, 8.777 **16.** 26.0999, 26.199, 25.99, 26.1909

17. LIBRARY Books in the library are placed on shelves in order according to their Dewey Decimal numbers. Arrange these numbers in order from least to greatest.

Book Number
943.678
943.6
943.67

18. ANALYZE TABLES The following table shows the amount of money Sonia spent on lunch each day this week. Order the amounts from least to greatest and then find the median amount she spent on lunch.

Day	Mon.	Tue.	Wed.	Thu.	Fri.
Amount Spent ($)	4.45	4.39	4.23	4.53	4.38

Copyright © Glencoe/McGraw-Hill, a division of The McGraw-Hill Companies, Inc.

NAME ______________________________ DATE ____________ PERIOD _____

3-3 Study Guide and Intervention

Rounding Decimals

To round a decimal, first underline the digit to be rounded. Then look at the digit to the right of the place being rounded.

- If the digit is 4 or less, the underlined digit remains the same.
- If the digit is 5 or greater, add 1 to the underlined digit.

Example 1 **Round 6.58 to the nearest tenth.**

Underline the digit to be rounded.	Look at the digit to the right of the underlined digit.	Since the digit to the right is 8, add one to the underlined digit.
$6.\underline{5}8$	$6.\underline{5}\mathbf{8}$	6.6

To the nearest tenth, 6.58 rounds to 6.6.

Example 2 **Round 86.943 to the nearest hundredth.**

Underline the digit to be rounded.	Look at the digit to the right of the underlined digit.	Since the digit is 3 and $3 < 5$, the digit 4 remains the same.
$86.9\underline{4}3$	$86.9\underline{4}\mathbf{3}$	86.94

To the nearest hundredth, 86.943 rounds to 86.94.

Exercises

Round each decimal to the indicated place-value position.

1. 3.21; tenths

2. 2.0505; thousandths

3. 6.5892; hundredths

4. 235.709; hundredths

5. 0.0914; thousandths

6. 34.35; tenths

7. 500.005; hundredths

8. 2.5134; tenths

9. 0.0052; thousandths

10. 0.0052; hundredths

11. 131.1555; thousandths

12. 232.88; tenths

Copyright © Glencoe/McGraw-Hill, a division of The McGraw-Hill Companies, Inc.

NAME ______________________ DATE ____________ PERIOD ______

3-3 Practice

Rounding Decimals

Round each decimal to the indicated place-value position.

1. 8.239; tenths

2. 3.666; tenths

3. 4.47; ones

4. 10.86; ones

5. 3.299; hundredths

6. 20.687; hundredths

7. 2.3654; thousandths

8. 69.0678; thousandths

9. 5.58214; hundredths

10. 468.09156; thousandths

11. $46.49; tens

12. 1,358.761; tens

13. LANGUAGES In the United States, about 1.64 million people speak French as their primary language. Round this number to the nearest million.

14. SHOPPING The price of a pound of cooked shrimp was $3.29. How much was this to the nearest dollar?

15. COMPUTERS Crystal has filled up 13.57 gigabytes of her computer's hard drive. Round this amount to the nearest tenth of a gigabyte.

16. CURRENCY Recently, one Canadian dollar was equal to 0.835125 U.S. dollars. Round this amount of U.S. dollars to the nearest cent.

CALCULATOR **A calculator will often show the results of a calculation with a very long decimal. Round each of the numbers on the calculator displays to the nearest thousandth.**

17. 35.67381216

18. 1342.409444

19. .5235728864

20. RACING The table shows the times for a canoe paddling race at summer camp. Will it help to round these times to the nearest tenth before listing them in in order from least to greatest? Explain.

Canoe Race

Team	Time (h)
Cougars	1.751
Moose	1.824
Jack Rabbits	1.665
Bears	1.739

Copyright © Glencoe/McGraw-Hill, a division of The McGraw-Hill Companies, Inc.

NAME ______________________ DATE ____________ PERIOD ______

3-4 Study Guide and Intervention

Estimating Sums and Differences

Estimation Methods	
Rounding	Estimate by rounding each decimal to the nearest whole number that is easy for you to add or subtract mentally.
Clustering	Estimate by rounding a group of close numbers to the same number.
Front-End Estimation	Estimate by adding or subtracting the values of the digits in the front place..

Example 1 **Estimate 14.07 + 43.22 using front-end estimation.**

Add the front digits.

$$\begin{array}{r} 14.07 \\ +\ 43.22 \\ \hline 5 \end{array}$$

Add the next digits.

$$\begin{array}{r} 14.07 \\ +\ 43.22 \\ \hline 57.00 \end{array}$$

An estimate for 14.07 + 43.22 is 57.

Example 2 **Use clustering to estimate $7.62 + $7.89 + $8.01 + $7.99.**

To use clustering, round each addend to the same number.

$$\begin{array}{rcr} 7.62 & \rightarrow & 8.00 \\ 7.89 & \rightarrow & 8.00 \\ 8.01 & \rightarrow & 8.00 \\ +\ 7.99 & \rightarrow & +\ 8.00 \\ & & \hline 32.00 \end{array}$$

An estimate for $7.62 + $7.89 + $8.01 + $7.99 is $32.

Exercises

Estimate using rounding.

1. 59.118 + 17.799 **2.** $45.85 + $6.82 **3.** 4.65 + 4.44

Estimate using clustering.

4. $0.99 + $1.15 + $0.52 **5.** 3.65 + 4.02 + 3.98 **6.** 6.87 + 6.97 + 7.39

Estimate using front-end estimation.

7. $\begin{array}{r} 81.23 \\ +\ 5.51 \\ \hline \end{array}$ **8.** $\begin{array}{r} 42.06 \\ +\ 17.39 \\ \hline \end{array}$ **9.** $\begin{array}{r} 754.23 \\ -\ 23.17 \\ \hline \end{array}$

Lesson 3–4

Copyright © Glencoe/McGraw-Hill, a division of The McGraw-Hill Companies, Inc.

NAME ______________________________ DATE __________ PERIOD _____

3-4 Practice

Estimating Sums and Differences

Estimate using rounding.

1. 68.99 + 22.31

2. 39.57 + 18.34

3. 81.25 − 23.16

4. 21.56 − 19.62

5. 5.69 + 3.47 + 8.02

6. 6.6 + 1.22 + 5.54

Estimate using clustering.

7. \$4.56 + \$4.79 + \$5.21 + \$5.38

8. 9.7325 + 9.55 + 10.333

9. 39.8 + 39.6 + 40.21 + 40.47

10. \$69.72 + \$70.44 + \$70.59 + \$69.56

Estimate using front-end estimation.

11. 34.87 − 29.12

12. 69.45 − 44.8

13. \$78.69 + \$31.49

14. \$258.32 + \$378.60

15. **SHOPPING** Miriam bought a basketball for \$24.99 and basketball shoes for \$47.79. About how much did Miriam spend on the ball and shoes?

16. **PRECIPITATION** Albuquerque gets an average of 6.35 inches of precipitation a year. Phoenix gets an average of 6.82 inches a year. About how many more inches of precipitation does Phoenix get than Albuquerque using rounding and using front-end estimation?

Copyright © Glencoe/McGraw-Hill, a division of The McGraw-Hill Companies, Inc.

NAME ______________________________ DATE ____________ PERIOD _____

3-5 Study Guide and Intervention

Adding and Subtracting Decimals

To add or subtract decimals, line up the decimal points then add or subtract digits in the same place-value position. Estimate first so you know if your answer is reasonable.

Example 1 **Find the sum of 61.32 + 8.26.**

First, estimate the sum using front-end estimation.

61.32 + 8.26 → 61 + 8 = 69

$$\begin{array}{r} 61.32 \\ +\ 8.26 \\ \hline 69.58 \end{array}$$

Since the estimate is close, the answer is reasonable.

Example 2 **Find 2.65 − 0.2.**

Estimate: 2.65 − 0.2 → 3 − 0 = 3

$$\begin{array}{r} 2.65 \\ -\ 0.20 \\ \hline 2.45 \end{array}$$

Annex a zero.

Since the estimate is close, the answer is reasonable.

Exercises

Find each sum or difference.

1. $\begin{array}{r} 2.3 \\ +\ 4.1 \\ \hline \end{array}$ **2.** $\begin{array}{r} \$13.67 \\ -\ 7.19 \\ \hline \end{array}$ **3.** $\begin{array}{r} 0.0123 \\ -\ 0.0028 \\ \hline \end{array}$ **4.** $\begin{array}{r} 132.346 \\ +\ 0.486 \\ \hline \end{array}$

5. $\begin{array}{r} 113.7999 \\ +\ 6.2001 \\ \hline \end{array}$ **6.** $\begin{array}{r} 0.0058 \\ -\ 0.0026 \\ \hline \end{array}$ **7.** $\begin{array}{r} \$5.63 \\ +\ 4.10 \\ \hline \end{array}$ **8.** $\begin{array}{r} 5.00921 \\ -4.00013 \\ \hline \end{array}$

9. 0.2 + 5.64 + 9.005 **10.** 12.36 − 4.081

11. 216.8 − 34.055 **12.** 4.62 + 3.415 + 2.4

Copyright © Glencoe/McGraw-Hill, a division of The McGraw-Hill Companies, Inc.

NAME ______________________ DATE ____________ PERIOD _____

3-5 Practice

Adding and Subtracting Decimals

Find each sum.

1. 5.4 + 6.5

2. 6.0 + 3.8

3. 3.65 + 4

4. 52.47 + 13.21

5. 91.64 + 19.5

6. 0.675 + 28

Find each difference.

7. 7.8 – 4.5

8. 69 – 12.88

9. 17.46 – 6.79

10. 74 – 59.29

11. 87.31 – 25.09

12. 19.75 – 12.98

ALGEBRA **Evaluate each expression if a = 219.6 and b = 12.024.**

13. $a - b$

14. $b + a$

15. $a - 13.45 - b$

Find the value of each expression.

16. $4.3 + 6 \times 7$

17. $3^2 - 2.55$

18. $19.7 - 4^2$

19. BIKE RIDING The table shows the distances the members of two teams rode their bicycles for charity.

a. How many total miles did Lori's team ride?

b. How many more miles did Lori's team ride than Tati's team?

Distances Ridden for Charity

Lori's Team		Tati's Team	
Lori	13.8 mi	Tati	13.6 mi
Marcus	11.8 mi	Luis	15.1 mi
Hassan	15.4 mi		

Copyright © Glencoe/McGraw-Hill, a division of The McGraw-Hill Companies, Inc.

NAME ______________________ DATE ____________ PERIOD _____

3-6 Study Guide and Intervention

Multiplying Decimals by Whole Numbers

When you multiply a decimal by a whole number, you multiply the numbers as if you were multiplying all whole numbers. Then you use estimation or you count the number of decimal places to decide where to place the decimal point. If there are not enough decimal places in the product, annex zeros to the left.

Example 1 **Find 6.25 × 5.**

Method 1 Use estimation.

Round 6.25 to 6.
$6.25 \times 5 \rightarrow 6 \times 5$ or 30

```
  1 2
  6.25
×    5
 31.25
```

Since the estimate is 30 place the decimal point after 31.

Method 2 Count decimal places.

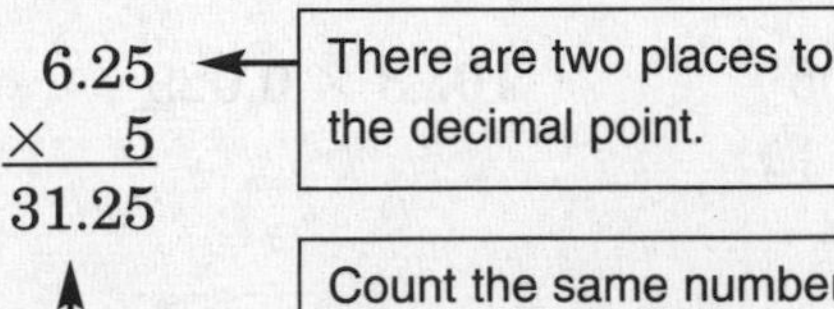


```
  6.25
×    5
 31.25
```

Count the same number of decimal places from right to left.

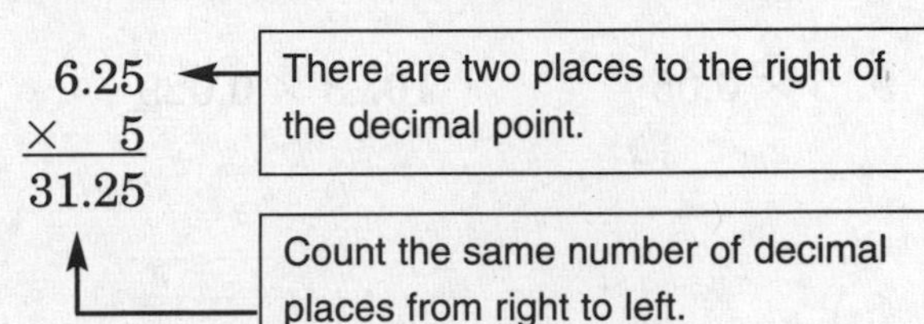

Example 2 **Find 3 × 0.0047.**

```
     2
0.0047
×    3
0.0141
```

There are four decimal places.

Annex a zero on the left of 141 to make four decimal places.

Example 3 **Find 6.3 × 1,000.**

Method 1 Use paper and pencil.

```
 1,000
 × 6.3
 3 000
60 000
6,300.0
```

Method 2 Use mental math.

Move the decimal point to the right the same number of zeros that are in 1,000 or 3 places.

$6.3 \times 1{,}000 = 6{,}300$

Exercises

Multiply.

1. 8.03 × 3 **2.** 6 × 12.6 **3.** 2 × 0.012 **4.** 0.0008 × 9

5. 2.32 × 10 **6.** 6.8 × 100 **7.** 5.2 × 1000 **8.** 1.412 × 100

Lesson 3–6

Copyright © Glencoe/McGraw-Hill, a division of The McGraw-Hill Companies, Inc.

NAME ______________________________ DATE ____________ PERIOD _____

3-6 Practice

Multiplying Decimals by Whole Numbers

Multiply.

1. 0.8×6 **2.** 0.7×4 **3.** 1.9×5 **4.** 3.4×9

5. 6×3.4 **6.** 5.2×9 **7.** 0.6×6 **8.** 4×0.8

9. 5×0.05 **10.** 3×0.029 **11.** 0.0027×15 **12.** 0.0186×92

ALGEBRA Evaluate each expression.

13. $5.02h$ if $h = 36$ **14.** $72.33j$ if $j = 3$ **15.** $21k$ if $k = 24.09$

Multiply.

16. 4.23×100 **17.** $3.7 \times 1{,}000$ **18.** 2.6×10 **19.** $4.2 \times 1{,}000$

20. 1.23×100 **21.** $5.14 \times 1{,}000$ **22.** 6.7×10 **23.** $7.89 \times 1{,}000$

24. **SHOPPING** Basketballs sell for $27.99 each at the Super D and for $21.59 each at the Bargain Spot. If the coach buys a dozen basketballs, how much can he save by buying them at the Bargain Spot? Justify your answer.

25. **SCHOOL** Jaimie purchases 10 pencils at the school bookstore. They cost $0.30 each. How much did she spend on pencils?

Copyright © Glencoe/McGraw-Hill, a division of The McGraw-Hill Companies, Inc.

NAME ______________________________ DATE ______________ PERIOD _____

3-7 Study Guide and Intervention

Multiplying Decimals

When you multiply a decimal by a decimal, multiply the numbers as if you were multiplying all whole numbers. To decide where to place the decimal point, find the sum of the number of decimal places in each factor. The product has the same number of decimal places.

Example 1 **Find 5.2 × 6.13.**

Estimate: 5 × 6 or 30

```
    5.2  ←— one decimal place
 × 6.13  ←— two decimal places
    156
    52
   312
 31.876  ←— three decimal places
```

The product is 31.876. Compared to the estimate, the product is reasonable.

Example 2 **Evaluate 0.023*t* if *t* = 2.3.**

$0.023t = 0.023 \times 2.3$ Replace *t* with 2.3.

```
  0.023  ←— three decimal places
  × 2.3  ←— one decimal place
     69
    46
 0.0529  ←— Annex a zero to make four decimal places.
```

Exercises

Multiply.

1. 7.2 × 2.1 **2.** 4.3 × 8.5 **3.** 2.64 × 1.4

4. 14.23 × 8.21 **5.** 5.01 × 11.6 **6.** 9.001 × 4.2

ALGEBRA Evaluate each expression if $x = 5.07$, $y = 1.5$, and $z = 0.403$.

7. $3.2x + y$ **8.** $yz + x$ **9.** $z \times 7.06 - y$

Copyright © Glencoe/McGraw-Hill, a division of The McGraw-Hill Companies, Inc.

3-7 Practice

Multiplying Decimals

Multiply.

1. 0.3×0.9 **2.** 2.6×1.7 **3.** 1.09×5.4 **4.** 17.2×12.86

5. 0.56×0.03 **6.** 4.9×0.02 **7.** 2.07×2.008 **8.** 26.02×2.006

ALGEBRA Evaluate each expression if $r = 0.034$, $s = 4.05$, and $t = 2.6$.

9. $5.027 + 4.68r$ **10.** $2.9s - 3.7t$ **11.** $4.13s + r$ **12.** rst

13. MINING A mine produces 42.5 tons of coal per hour. How much coal will the mine produce in 9.5 hours?

14. SHOPPING Ms. Morgan bought 3.5 pounds of bananas at $0.51 a pound and 4.5 pounds of pineapple at $1.19 a pound. How much did she pay for the bananas and pineapple?

Copyright © Glencoe/McGraw-Hill, a division of The McGraw-Hill Companies, Inc.

NAME ______________________ DATE ____________ PERIOD ______

3-8 Study Guide and Intervention

Dividing Decimals by Whole Numbers

When you divide a decimal by a whole number, place the decimal point in the quotient above the decimal point in the dividend. Then divide as you do with whole numbers.

Example 1 **Find 8.73 ÷ 9.**

Estimate: 9 ÷ 9 = 1

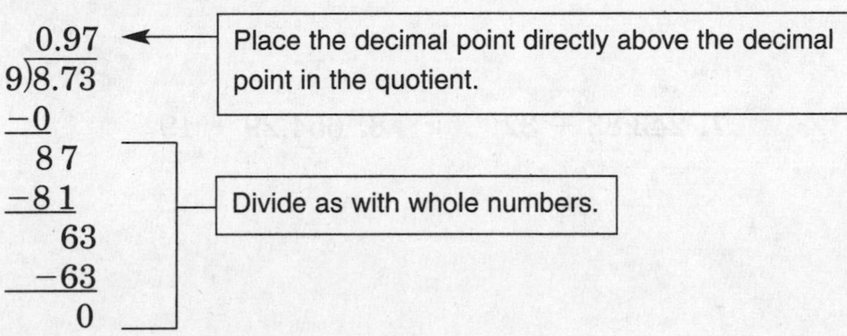

8.73 ÷ 9 = 0.97 Compared to the estimate, the quotient is reasonable.

Example 2 **Find 8.58 ÷ 12.**

Estimate: 10 ÷ 10 = 1

```
   0.715   ← Place the decimal point.
12)8.580
  −84
    18
   −12       Annex a zero to continue dividing.
    60
   −60
     0
```

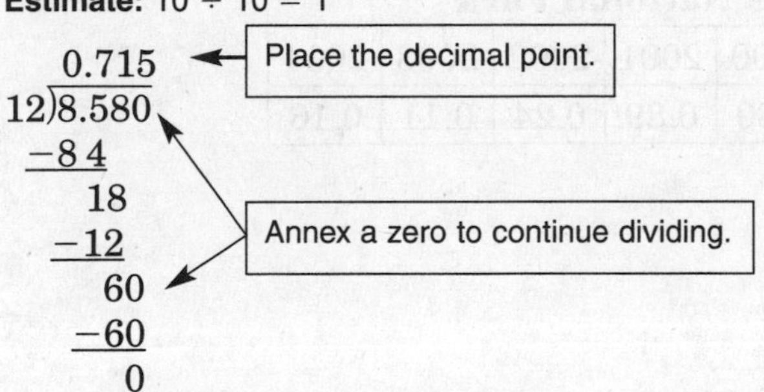

8.58 ÷ 12 = 0.715 Compared to the estimate, the quotient is reasonable.

Exercises

Divide.

1. 9.2 ÷ 4 **2.** 4.5 ÷ 5 **3.** 8.6 ÷ 2 **4.** 2.89 ÷ 4

5. 3.2 ÷ 4 **6.** 7.2 ÷ 3 **7.** 7.5 ÷ 5 **8.** 3.25 ÷ 5

Lesson 3–8

Copyright © Glencoe/McGraw-Hill, a division of The McGraw-Hill Companies, Inc.

NAME ______________________ DATE __________ PERIOD _____

3-8 Practice

Dividing Decimals by Whole Numbers

Divide. Round to the nearest tenth if necessary.

1. 25.2 ÷ 4 **2.** 147.2 ÷ 8 **3.** 5.69 ÷ 7 **4.** 13.28 ÷ 3

5. 22.5 ÷ 15 **6.** 65.28 ÷ 12 **7.** 243.83 ÷ 32 **8.** 654.29 ÷ 19

9. WEATHER What is the average January precipitation in Arches National Park? Round to the nearest hundredth if necessary.

January Precipitation in Arches National Park								
Year	1997	1998	1999	2000	2001	2002	2003	2004
Precipitation (in.)	1.09	0.013	0.54	0.80	0.89	0.24	0.11	0.16

Source: National Park Service

10. SHOPPING A 3-pack of boxes of juice costs $1.09. A 12-pack of boxes costs $4.39. A case of 24 boxes costs $8.79. Which is the best buy? Explain your reasoning.

Copyright © Glencoe/McGraw-Hill, a division of The McGraw-Hill Companies, Inc.

NAME ______________________________ DATE ____________ PERIOD _____

3-9 Study Guide and Intervention

Dividing by Decimals

When you divide a decimal by a decimal, multiply both the divisor and the dividend by the same power of ten. Then divide as with whole numbers.

Example 1 **Find 10.14 ÷ 5.2.**

Estimate: 10 ÷ 5 = 2

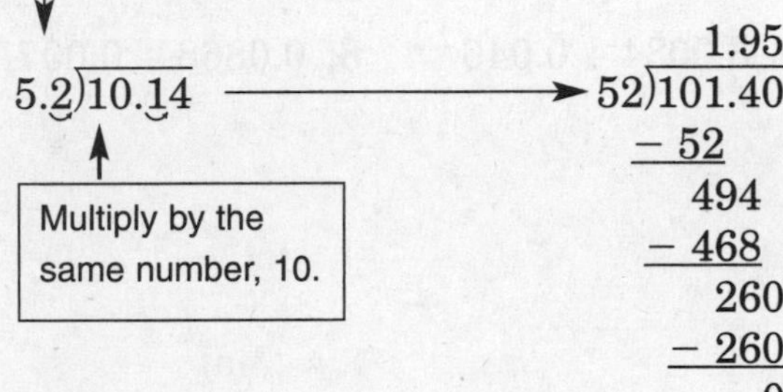


```
      1.95
 52)101.40     Place the decimal point.
                Divide as with whole numbers.
   - 52
     494
   - 468
      260      Annex a zero to continue.
    - 260
        0
```

10.14 divided by 5.2 is 1.95. Compare to the estimate.

Check: 1.95 × 5.2 = 10.14 ✓

Example 2 **Find 4.09 ÷ 0.02.**

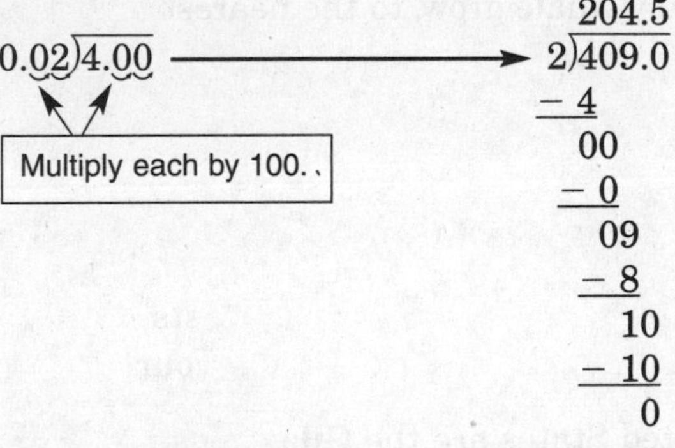


```
     204.5
  2)409.0     Place the decimal point.
               Divide.
   - 4
     00
    - 0
     09
    - 8
     10       Write a zero in the dividend
   - 10       and continue to divide.
      0
```

4.09 ÷ 0.02 is 204.5.

Check: 204.5 × 0.02 = 4.09 ✓

Exercises

Divide.

1. 9.8 ÷ 1.4

2. 4.41 ÷ 2.1

3. 16.848 ÷ 0.72

4. 8.652 ÷ 1.2

5. 0.5 ÷ 0.001

6. 9.594 ÷ 0.06

Lesson 3–9

Copyright © Glencoe/McGraw-Hill, a division of The McGraw-Hill Companies, Inc.

3-9 Practice

Dividing by Decimals

Divide.

1. 12.92 ÷ 3.4 **2.** 22.47 ÷ 0.7 **3.** 0.025 ÷ 0.5 **4.** 7.224 ÷ 0.08

5. 0.855 ÷ 9.5 **6.** 0.9 ÷ 0.12 **7.** 3.0084 ÷ 0.046 **8.** 0.0868 ÷ 0.007

9. **WHALES** After its first day of life, a baby blue whale started growing. It grew 47.075 inches. If the average baby blue whale grows at a rate of 1.5 inches a day, for how many days did the baby whale grow, to the nearest tenth of a day?

10. **LIZARDS** The two largest lizards in the United States are the Gila Monster and the Chuckwalla. The average Gila Monster is 0.608 meter long. The average Chuckwalla is 0.395 meters long. How many times longer is the Gila Monster than the Chuckwalla to the nearest hundredth?

Copyright © Glencoe/McGraw-Hill, a division of The McGraw-Hill Companies, Inc.

3-10 Study Guide and Intervention

Problem-Solving Investigation: Reasonable Answers

When solving problems, one strategy that is helpful is to *determine reasonable answers.* If you are solving a problem with big numbers, or a problem with information that you are unfamiliar with, it may be helpful to look back at your answer to determine if it is reasonable.

You can use the *determine reasonable answers* strategy, along with the following four-step problem solving plan to solve a problem.

1 Understand – Read and get a general understanding of the problem.

2 Plan – Make a plan to solve the problem and estimate the solution.

3 Solve – Use your plan to solve the problem.

4 Check – Check the reasonableness of your solution.

Example ANIMALS **The average height of a male chimpanzee is 1.2 meters, and the average height of a female chimpanzee is 1.1 meters. What is a reasonable height in feet of a male chimpanzee?**

Understand We know the average height in meters of a male chimpanzee.

We need to find a reasonable height in feet.

Plan One meter is very close to one yard. One yard is equal to 3 feet. So, estimate how many feet would be in 1.2 yards.

Solve 1.2 yards would be more than 3 feet, but less than 6 feet.

So, a reasonable average height of a male chimpanzee is about 4 feet.

Check Since 1.2 yd = 3.6 ft, the answer of 4 feet is reasonable.

Exercise

SHOPPING Alexis wants to buy 2 bracelets for $6.95 each, 1 pair of earrings for $4.99, and 2 necklaces for $8.95 each. Does she need $40 or will $35 be more reasonable? Explain.

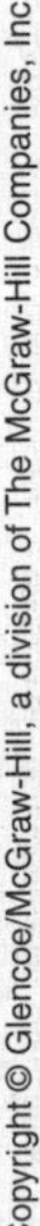

Copyright © Glencoe/McGraw-Hill, a division of The McGraw-Hill Companies, Inc.

Lesson 3–10

NAME ______________________ DATE __________ PERIOD ______

3-10 Practice

Problem-Solving Investigation: Reasonable Answers

Mixed Problem Solving

Use the determine reasonable answers strategy to solve Exercises 1 and 2.

1. **LIFE EXPECTANCY** Use the graph below to determine whether 80, 85, or 90 years is a reasonable prediction of the life expectancy of a person born in 2020.

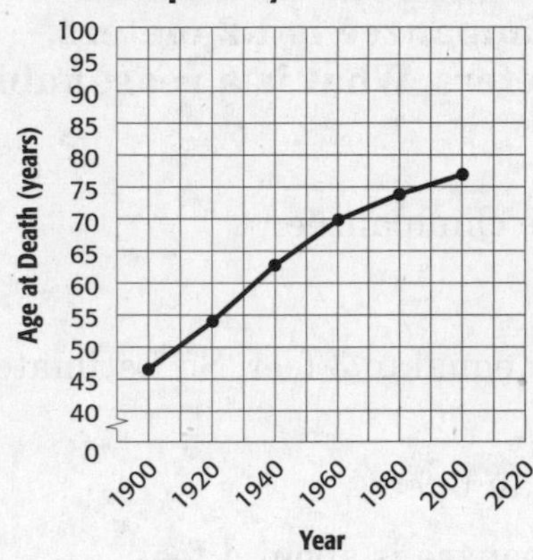

2. **SNACKS** Paolo is stocking up on after-school snacks. He wants to buy 2 pounds of bananas at $0.79 per pound, 2 cans of mixed nuts at $3.89 a can, and a bottle of apple juice at $1.19 a bottle. Does he need to bring $20 to the store or will $15 be enough? Explain your reasoning.

Use any strategy to solve Exercises 3–6. Some strategies are shown below.

Problem-Solving Strategies
• Solve a simpler problem.
• Draw a diagram.
• Determine reasonable answers.

3. **CARVINGS** In how many ways can Kwan line up her carvings of a duck, a gull, and a pelican on a shelf?

4. **CARNIVAL** There are 56 students in the sixth grade. Ms. Rockwell's class is sponsoring a carnival for the sixth graders at the school. The class has spent $40 on decorations and $10 on publicity. To pay for the expenses, an entrance fee of $0.75 is being considered. Is this a reasonable amount to charge?

5. **PARKS** The four largest national parks in the United States are in Alaska. The largest is Wrangell-St. Elias at 8.3 million acres. The fourth largest is Katmai at 1.48 million acres. How many times larger is Wrangell-St. Elias than Katmai to the nearest tenth million?

6. **RACING** Hector ran in the city charity race for four years. His times in minutes were: 14.8, 22.3, 26.7, and 31.9. What was his mean time for the four years to the nearest tenth minute?

Copyright © Glencoe/McGraw-Hill, a division of The McGraw-Hill Companies, Inc.

NAME ______________________ DATE ____________ PERIOD _____

4-1 Study Guide and Intervention

Greatest Common Factor

The **greatest common factor (GCF)** of two or more numbers is the greatest of the common factors of the numbers. To find the GCF, you can make a list or use prime factors.

Example 1 **Find the GCF of 12 and 30.**

Make an organized list of the factors for each number.

12: 1×12, 2×6, 3×4
30: 1×30, 2×15, 3×10, 5×6

The common factors are 1, 2, 3, and 6. The greatest is 6. The GCF of 12 and 30 is 6.

Example 2 **Find the GCF of 18 and 27 by using prime factors.**

Write the prime factorizations of 18 and 27.

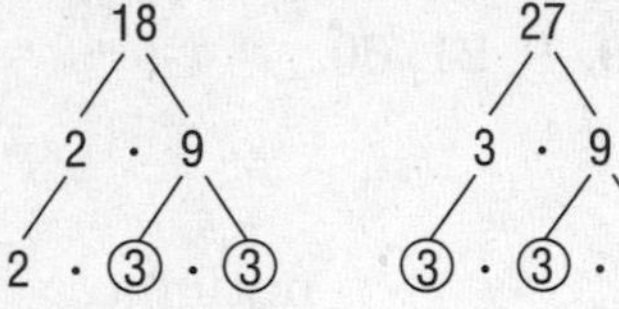

The common prime factors are 3 and 3. So, the GCF of 18 and 27 is 3×3 or 9.

Exercises

Find the GCF of each set of numbers by making a list.

1. 8 and 12 **2.** 10 and 15 **3.** 81 and 27

Find the GCF of each set of numbers by using prime factors.

4. 15 and 20 **5.** 6 and 12 **6.** 28 and 42

Find the GCF of each set of numbers.

7. 21 and 9 **8.** 15 and 7 **9.** 54 and 81

10. 30 and 45 **11.** 44 and 55 **12.** 35, 20, and 15

Copyright © Glencoe/McGraw-Hill, a division of The McGraw-Hill Companies, Inc.

NAME ______________________ DATE __________ PERIOD ______

4-1 Practice

Greatest Common Factor

Identify the common factors of each set of numbers.

1. 12 and 20

2. 12, 24, 36

3. 15, 33, 45

Find the GCF of each set of numbers.

4. 12 and 30

5. 50 and 40

6. 20 and 27

7. 28, 42, 56

8. 14, 56, 63

9. 9, 21, 60

Find three numbers whose GCF is the indicated value.

10. 3

11. 16

12. 18

TOYS For Exercises 13 and 14, use the following information.

A store is organizing toys into bins. The toys must be put into bins such that each bin contains the same number of toys without mixing the toys.

Toys to Place in Bins	
Toy	**Number of Toys**
airplanes	36
boats	72
cars	60

13. What is the greatest number of toys that can be put in a bin?

14. How many bins are needed for each type of toy?

Copyright © Glencoe/McGraw-Hill, a division of The McGraw-Hill Companies, Inc.

NAME ______________________ DATE __________ PERIOD _____

4-2 Study Guide and Intervention

Simplifying Fractions

Fractions that have the same value are **equivalent fractions**. To find equivalent fractions, you can multiply or divide the numerator and denominator by the same nonzero number.

Example 1 **Replace the ● with a number so that $\frac{1}{2} = \frac{●}{10}$.**

Since 2 × 5 = 10, multiply the numerator and denominator by 5.

$$\frac{1}{2} = \frac{●}{10} \quad (\times 5) \qquad \frac{1}{2} = \frac{5}{10}. \quad (\times 5)$$

When the GCF of the numerator and denominator is 1, the fraction is in simplest form. To write a fraction in simplest form, you can divide the numerator and denominator by the GCF.

Example 2 **Write $\frac{12}{30}$ in simplest form.**

The GCF of 12 and 30 is 6.

$$\frac{12}{30} = \frac{2}{5} \quad (\div 6)$$

Divide the numerator and denominator by the GCF, 6.

The GCF of 2 and 5 is 1, so $\frac{2}{5}$ is in simplest form.

Exercises

Replace each ● with a number so the fractions are equivalent.

1. $\frac{1}{5} = \frac{●}{15}$
2. $\frac{12}{18} = \frac{2}{●}$
3. $\frac{●}{14} = \frac{27}{42}$

Write each fraction in simplest form. If the fraction is already in simplest form, write *simplest form*.

4. $\frac{6}{30}$
5. $\frac{2}{3}$
6. $\frac{6}{8}$
7. $\frac{21}{28}$
8. $\frac{15}{30}$
9. $\frac{7}{10}$

Copyright © Glencoe/McGraw-Hill, a division of The McGraw-Hill Companies, Inc.

Lesson 4–2

NAME ______________________ DATE __________ PERIOD _____

4-2 Practice

Simplifying Fractions

Replace each ● with a number so the fractions are equivalent.

1. $\frac{1}{3} = \frac{●}{9}$ **2.** $\frac{1}{4} = \frac{●}{16}$ **3.** $\frac{●}{2} = \frac{8}{16}$ **4.** $\frac{●}{8} = \frac{9}{24}$

5. $\frac{1}{2} = \frac{16}{●}$ **6.** $\frac{12}{21} = \frac{4}{●}$ **7.** $\frac{30}{36} = \frac{●}{6}$ **8.** $\frac{28}{42} = \frac{●}{3}$

Write each fraction in simplest form. If the fraction is already in simplest form, write *simplest form.*

9. $\frac{7}{28}$ **10.** $\frac{9}{15}$ **11.** $\frac{10}{42}$

12. $\frac{12}{42}$ **13.** $\frac{17}{28}$ **14.** $\frac{24}{64}$

Write two fractions that are equivalent to the given fraction.

15. $\frac{3}{10}$ **16.** $\frac{7}{13}$ **17.** $\frac{15}{33}$

18. **ANIMALS** In Ms Reyes' class, 4 out of the 30 students had guinea pigs as pets. Express this fraction in simplest form.

19. **ANALYZE GRAPHS** The bar graph shows the number of titles held by the top seven women Wimbledon tennis champions. In simplest form, what fraction of the number of titles is held by Steffi Graf?

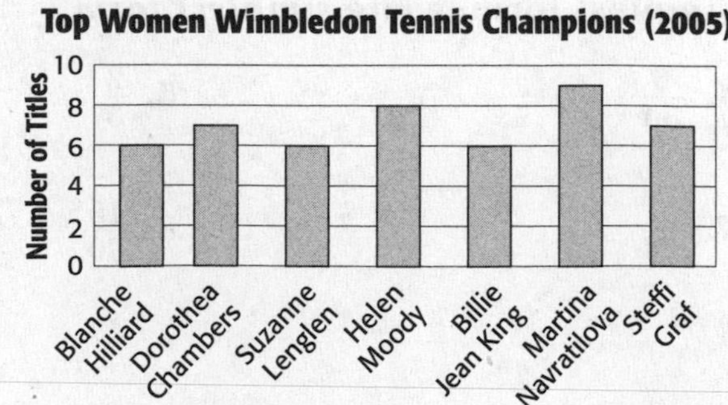

Copyright © Glencoe/McGraw-Hill, a division of The McGraw-Hill Companies, Inc.

NAME ______________________________ DATE ____________ PERIOD ______

4-3 Study Guide and Intervention

Mixed Numbers and Improper Fractions

The number $2\frac{2}{3}$ is a mixed number. A **mixed number** indicates the sum of a whole number and a fraction. The number $\frac{5}{3}$ is an improper fraction. **Improper fractions** have values that are greater than or equal to 1. Mixed numbers can be written as mixed numbers or as improper fractions.

Example 1 **Write $2\frac{1}{3}$ as an improper fraction.**

$2\frac{1}{3} \rightarrow 2 \times \frac{3}{3} + \frac{1}{3} = \frac{7}{3}$ Think: $2 \times 3 = 6$ and $6 + 1 = 7$

Check: Use a model.

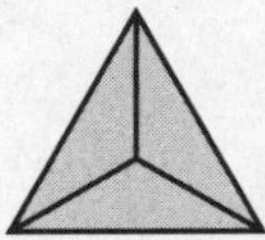 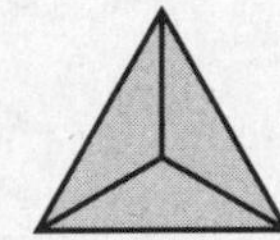 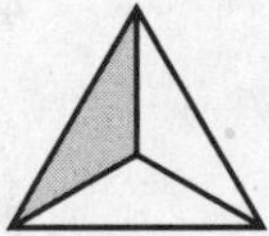

$\frac{7}{3} = 2 + \frac{1}{3}$ or $2\frac{1}{3}$ ✓

Example 2 **Write $\frac{9}{4}$ as a mixed number.**

Divide 9 by 4. Use the remainder as the numerator of the fraction.

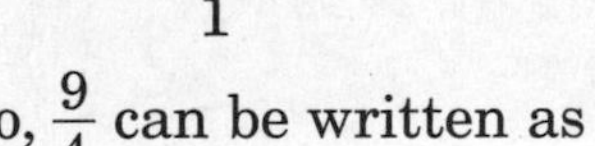

$$\begin{array}{r} 2\frac{1}{4} \\ 4\overline{)9} \\ \underline{-8} \\ 1 \end{array}$$

So, $\frac{9}{4}$ can be written as $2\frac{1}{4}$.

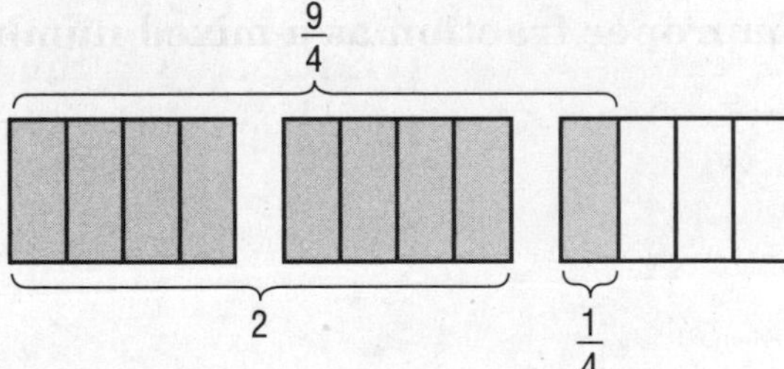

Exercises

Write each mixed number as an improper fraction.

1. $3\frac{1}{8}$ **2.** $2\frac{4}{5}$ **3.** $2\frac{1}{2}$ **4.** $1\frac{2}{3}$

5. $2\frac{1}{9}$ **6.** $3\frac{7}{10}$ **7.** $2\frac{3}{8}$ **8.** $1\frac{3}{4}$

Write each improper fraction as a mixed number or a whole number.

9. $\frac{7}{4}$ **10.** $\frac{5}{3}$ **11.** $\frac{3}{2}$ **12.** $\frac{11}{8}$

13. $\frac{22}{5}$ **14.** $2\frac{15}{15}$ **15.** $\frac{25}{4}$ **16.** $\frac{16}{3}$

Copyright © Glencoe/McGraw-Hill, a division of The McGraw-Hill Companies, Inc.

NAME ______________________ DATE ____________ PERIOD _____

4-3 Practice

Mixed Numbers and Improper Fractions

Write each mixed number as an improper fraction.

1. $4\frac{2}{3}$ **2.** $2\frac{1}{2}$ **3.** $5\frac{3}{7}$ **4.** $3\frac{5}{6}$

5. $6\frac{1}{4}$ **6.** $5\frac{3}{5}$ **7.** $8\frac{1}{9}$ **8.** $6\frac{3}{4}$

9. SNAKES The garden snake that Fumiko measured was $7\frac{3}{4}$ inches long. Write the length as an improper fraction.

10. Express *four and seven eighths* as an improper fraction.

Write each improper fraction as a mixed number or a whole number.

11. $\frac{13}{4}$ **12.** $\frac{11}{10}$ **13.** $\frac{10}{3}$

14. $\frac{23}{7}$ **15.** $6\frac{14}{14}$ **16.** $\frac{8}{8}$

17. TREES A nursery is growing trees. Find the height of each tree in terms of feet. Write your answer as a mixed number in simplest form.

Trees in Nursery	
Tree	**Height (in.)**
Apricot	73
Peach	62
Pear	54
Plum	68

Copyright © Glencoe/McGraw-Hill, a division of The McGraw-Hill Companies, Inc.

NAME ______________________ DATE ____________ PERIOD ______

4-4 Study Guide and Intervention

Problem-Solving Investigation: Make an Organized List

When solving problems, one strategy that is helpful is to *make an organized list.* A list of all the possible combinations based on the information in the problem will help you solve the problem.

You can use the *make an organized list* strategy, along with the following four-step problem solving plan to solve a problem.

1 Understand – Read and get a general understanding of the problem.

2 Plan – Make a plan to solve the problem and estimate the solution.

3 Solve – Use your plan to solve the problem.

4 Check – Check the reasonableness of your solution.

Example 1 **ELECTIONS Tyler, McKayla, and Kareem are running for student council office. The three positions they could be elected for are president, treasurer, and secretary. How many possible ways could the three of them be elected?**

Understand You know that there are three positions and three students to fill the positions. You need to know the number of possible arrangements for them to be elected.

Plan Make a list of all the different possible arrangements. Use T for Tyler, M for McKayla, and K for Kareem.

Solve

President	T	T	K	K	M	M
Treasurer	M	K	M	T	T	K
Secretary	K	M	T	M	K	T

Check Check the answer by seeing if each student is accounted for in each situation.

Exercise

SHOPPING Khuan has to stop by the photo store, the gas station, the grocery store, and his grandmother's house. How many different ways can Khuan make the stops?

Lesson 4–4

Copyright © Glencoe/McGraw-Hill, a division of The McGraw-Hill Companies, Inc.

NAME ______________________ DATE ____________ PERIOD _____

4-4 Practice

Problem-Solving Investigation: Make an Organized List

Mixed Problem Solving

Use the make an organized list strategy to solve Exercises 1 and 2.

1. **FLAGS** Randy wants to place the flag of each of 3 countries in a row on the wall for an international fair. How many arrangements are possible?

2. **KITES** A store sells animal kites, box kites, and diamond kites in four different colors. How many combinations of kite type and color are possible?

Use any strategy to solve Exercises 3–7. Some strategies are shown below.

Problem-Solving Strategies
• Make a table.
• Guess and check.

3. **SHIRTS** A mail-order company sells 4 styles of shirts in 6 different colors. How many combinations of style and color are possible?

4. **PATTERNS** If the pattern continues, how many small squares are in the fifth figure of this pattern?

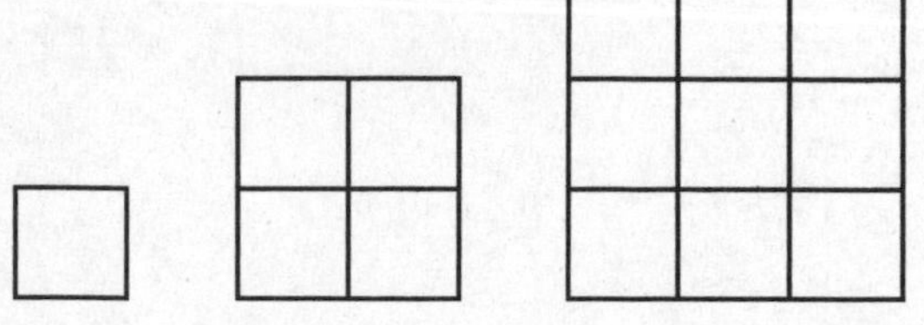

5. **FOOD** Is $6 enough money to buy a head of lettuce for $0.99, two pounds of tomatoes for $2.38, and two pounds of avocados for $2.78?

6. **MONEY** Nikki earns $45 a week pet sitting. How much does she earn each year?

7. **WRITING** The number of magazine articles Nora sold in her first four years is shown. At this rate, how many articles will she sell in the fifth year?

Year	Number Sold
1	2
2	4
3	7
4	11
5	?

Copyright © Glencoe/McGraw-Hill, a division of The McGraw-Hill Companies, Inc.

NAME ______________________________ DATE ______________ PERIOD _____

4-5 Study Guide and Intervention

Least Common Multiple

A **multiple** of a number is the product of the number and any counting number. The multiples of 2 are below.

$1 \times 2 = 2$ $2 \times 2 = 4$ $3 \times 2 = 6$ $4 \times 2 = 8$ $5 \times 2 = 10$

The smallest number that is a multiple of two or more whole numbers is the **least common multiple (LCM)** of the numbers.

Example 1 **Identify the first three common multiples of 3 and 6.**

Step 1 List the multiples of each number.

multiples of 3: 3, 6, 9, 12, 15, 18, ...

multiples of 6: 6, 12, 18, 24, 30, ...

Step 2 Identify the first three common multiples from the list.

The first three common multiples of 3 and 6 are 6, 12, and 18.

Example 2 **Find the LCM of 6 and 15 by using prime factors.**

Step 1 Write the prime factorization of each number.

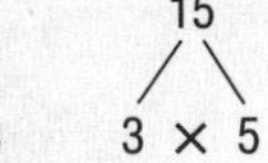
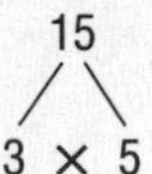

Step 2 Identify all common prime factors.

$6 = 2 \times 3$

$15 = \quad 3 \times 5$

Step 3 Find the product of all of the prime factors using each common prime factor once and any remaining factors.

The LCM is $2 \times 3 \times 5$ or 30.

Exercises

Identify the first three common multiples of each set of numbers.

1. 2 and 4 **2.** 5 and 10 **3.** 2 and 7

Find the LCM of each set of numbers.

4. 5 and 6 **5.** 6 and 9 **6.** 4 and 10

7. 9 and 27 **8.** 4 and 6 **9.** 5 and 7

Lesson 4–5

Copyright © Glencoe/McGraw-Hill, a division of The McGraw-Hill Companies, Inc.

NAME ______________________ DATE ____________ PERIOD ______

4-5 Practice

Least Common Multiple

Identify the first three common multiples of each set of numbers.

1. 4 and 5

2. 1 and 9

3. 3 and 4

4. 4, 6, and 8

Find the LCM of each set of numbers.

5. 3 and 5

6. 8 and 12

7. 3, 5, and 6

8. 6, 12, and 15

9. **PATTERNS** List the next four common multiples after the LCM of 3 and 8.

10. **E-MAIL** Alberto gets newsletters by e-mail. He gets one for sports every 5 days, one for model railroads every 10 days, and one for music every 8 days. If he got all three today, how many more days will it be until he gets all three newsletters on the same day?

Copyright © Glencoe/McGraw-Hill, a division of The McGraw-Hill Companies, Inc.

NAME ______________________ DATE __________ PERIOD ____

4-6 Study Guide and Intervention

Comparing and Ordering Fractions

To compare two fractions,

- Find the **least common denominator (LCD)** of the fractions; that is, find the least common multiple of the denominators.
- Write an equivalent fraction for each fraction using the LCD.
- Compare the numerators.

Example 1 **Replace ● with <, >, or = to make $\frac{1}{3}$ ● $\frac{5}{12}$ true.**

- The LCM of 3 and 12 is 12. So, the LCD is 12.
- Rewrite each fraction with a denominator of 12.

$\frac{1}{3} = \frac{●}{12}$ (×4), so $\frac{1}{3} = \frac{4}{12}$. $\frac{5}{12} = \frac{5}{12}$

- Now, compare. Since 4 < 5, $\frac{4}{12} < \frac{5}{12}$. So $\frac{1}{3} < \frac{5}{12}$.

Example 2 **Order $\frac{1}{6}$, $\frac{2}{3}$, $\frac{1}{4}$, and $\frac{3}{8}$ from least to greatest.**

The LCD of the fractions is 24. So, rewrite each fraction with a denominator of 24.

$\frac{1}{6} = \frac{●}{24}$ (×4), so $\frac{1}{6} = \frac{4}{24}$. $\frac{2}{3} = \frac{●}{24}$ (×8), so $\frac{2}{3} = \frac{16}{24}$.

$\frac{1}{4} = \frac{●}{24}$ (×6), so $\frac{1}{4} = \frac{6}{24}$. $\frac{3}{8} = \frac{●}{24}$ (×3), so $\frac{3}{8} = \frac{9}{24}$.

The order of the fractions from least to greatest is $\frac{1}{6}$, $\frac{1}{4}$, $\frac{3}{8}$, $\frac{2}{3}$.

Exercises

Replace each ● with <, >, or = to make a true sentence.

1. $\frac{5}{12}$ ● $\frac{3}{8}$
2. $\frac{6}{8}$ ● $\frac{3}{4}$
3. $\frac{2}{7}$ ● $\frac{1}{6}$

Order the fractions from least to greatest.

4. $\frac{3}{4}$, $\frac{3}{8}$, $\frac{1}{2}$, $\frac{1}{4}$
5. $\frac{2}{3}$, $\frac{1}{6}$, $\frac{5}{18}$, $\frac{7}{9}$
6. $\frac{1}{2}$, $\frac{5}{6}$, $\frac{5}{8}$, $\frac{5}{12}$

Copyright © Glencoe/McGraw-Hill, a division of The McGraw-Hill Companies, Inc.

NAME ______________________ DATE __________ PERIOD _____

4-6 Practice

Comparing and Ordering Fractions

Replace each ● with <, >, or = to make a true statement.

1. $\frac{11}{21}$ ● $\frac{2}{3}$

2. $\frac{1}{2}$ ● $\frac{9}{18}$

3. $2\frac{3}{8}$ ● $2\frac{8}{24}$

4. $6\frac{2}{3}$ ● $6\frac{12}{15}$

5. $5\frac{3}{4}$ ● $5\frac{8}{12}$

6. $\frac{2}{3}$ ● $\frac{10}{18}$

7. $\frac{18}{14}$ ● $1\frac{2}{7}$

8. $\frac{11}{12}$ ● $2\frac{1}{3}$

9. $\frac{34}{18}$ ● $1\frac{5}{6}$

Order the fractions from least to greatest.

10. $\frac{3}{5}, \frac{1}{4}, \frac{1}{2}, \frac{2}{5}$

11. $\frac{7}{9}, \frac{13}{18}, \frac{5}{6}, \frac{2}{3}$

12. $6\frac{3}{4}, 6\frac{1}{2}, 6\frac{5}{6}, 6\frac{3}{8}$

13. $2\frac{2}{3}, 2\frac{6}{15}, 2\frac{3}{5}, 2\frac{4}{9}$

14. MUSIC Ramundus is making a xylophone. So far, he has bars that are $1\frac{3}{4}$ feet, $1\frac{7}{12}$ feet, and $1\frac{2}{3}$ feet long. What is the length of the longest bar?

15. DANCE Alana practiced dancing for $\frac{11}{4}$ hours on Monday, $\frac{19}{8}$ hours on Wednesday, and $2\frac{3}{5}$ hours on Friday. On which day did she practice the closest to 2 hours? Explain your reasoning.

Copyright © Glencoe/McGraw-Hill, a division of The McGraw-Hill Companies, Inc.

NAME ______________________ DATE ____________ PERIOD _____

4-7 Study Guide and Intervention

Writing Decimals as Fractions

Decimals like 0.58, 0.12, and 0.08 can be written as fractions.

To write a decimal as a fraction, you can follow these steps.

1. Identify the place value of the last decimal place.
2. Write the decimal as a fraction using the place value as the denominator.

Example 1 **Write 0.5 as a fraction in simplest form.**

$0.5 = \frac{5}{10}$ 0.5 means five tenths.

$= \frac{\overset{1}{\cancel{5}}}{\underset{2}{\cancel{10}}}$ Simplify. Divide the numerator and denominator by the GCF, 5.

$= \frac{1}{2}$ So, in simplest form, 0.5 is $\frac{1}{2}$.

Example 2 **Write 0.35 as a fraction in simplest form.**

$0.35 = \frac{35}{100}$ 0.35 means 35 hundredths.

$= \frac{\overset{7}{\cancel{35}}}{\underset{20}{\cancel{100}}}$ Simplify. Divide the numerator and denominator by the GCF, 5.

$= \frac{7}{20}$ So, in simplest form, 0.35 is $\frac{7}{20}$.

Example 3 **Write 4.375 as a mixed number in simplest form.**

$4.375 = 4\frac{375}{1{,}000}$ 0.375 means 375 thousandths.

$= 4\frac{\overset{3}{\cancel{375}}}{\underset{8}{\cancel{1{,}000}}}$ Simplify. Divide by the GCF, 125.

$= 4\frac{3}{8}$

Exercises

Write each decimal as a fraction or mixed number in simplest form.

1. 0.9 **2.** 0.8 **3.** 0.27 **4.** 0.75

5. 0.34 **6.** 0.125 **7.** 0.035 **8.** 0.008

9. 1.4 **10.** 3.6 **11.** 6.28 **12.** 2.65

13. 12.05 **14.** 4.004 **15.** 23.205 **16.** 51.724

Lesson 4–7

Copyright © Glencoe/McGraw-Hill, a division of The McGraw-Hill Companies, Inc.

NAME ______________________ DATE __________ PERIOD _____

4-7 Practice

Writing Decimals as Fractions

Write each decimal as a fraction in simplest form.

1. 0.5

2. 0.8

3. 0.9

4. 0.75

5. 0.48

6. 0.72

7. 0.625

8. 0.065

9. 0.002

Write each decimal as a mixed number in simplest form.

10. 3.6

11. 10.4

12. 2.11

13. 29.15

14. 7.202

15. 23.535

16. DISTANCE The library is 0.96 mile away from Theo's home. Write this distance as a fraction in simplest form.

17. INSECTS A Japanese beetle has a length between 0.3 and 0.5 inch. Find two lengths that are within the given span. Write them as fractions in simplest form.

Copyright © Glencoe/McGraw-Hill, a division of The McGraw-Hill Companies, Inc.

NAME ______________________ DATE ____________ PERIOD _____

4-8 Study Guide and Intervention

Writing Fractions as Decimals

Fractions whose denominators are factors of 10, 100, or 1,000 can be written as decimals using equivalent fractions. Any fraction can also be written as a decimal by dividing the numerator by the denominator.

Example 1 **Write $\frac{3}{5}$ as a decimal.**

Since 5 is a factor of 10, write an equivalent fraction with a denominator of 10.

$$\frac{3 \times 2}{5 \times 2} = \frac{6}{10}$$

$$= 0.6$$

Therefore, $\frac{3}{5} = 0.6$.

Example 2 **Write $\frac{3}{8}$ as a decimal.**

Divide.

$$\begin{array}{r} 0.375 \\ 8\overline{)3.000} \\ -2\,4 \\ \hline 60 \\ -56 \\ \hline 40 \\ -40 \\ \hline 0 \end{array}$$

Therefore, $\frac{3}{8} = 0.375$.

Exercises

Write each fraction or mixed number as a decimal.

1. $\frac{3}{10}$ **2.** $\frac{3}{4}$ **3.** $\frac{1}{4}$ **4.** $\frac{3}{5}$

5. $\frac{1}{8}$ **6.** $2\frac{1}{4}$ **7.** $\frac{6}{20}$ **8.** $\frac{9}{25}$

9. $1\frac{3}{8}$ **10.** $1\frac{5}{8}$ **11.** $3\frac{5}{16}$ **12.** $4\frac{9}{20}$

Copyright © Glencoe/McGraw-Hill, a division of The McGraw-Hill Companies, Inc.

NAME ______________________________ DATE ______________ PERIOD ________

4-8 Practice

Writing Fractions as Decimals

Write each fraction or mixed number as a decimal.

1. $\frac{4}{5}$

2. $\frac{7}{20}$

3. $\frac{13}{250}$

4. $\frac{7}{8}$

5. $\frac{3}{16}$

6. $\frac{11}{32}$

7. $9\frac{29}{40}$

8. $7\frac{29}{80}$

9. $4\frac{11}{32}$

Replace each ● with <, >, or = to make a true sentence.

10. $\frac{1}{4}$ ● 0.2

11. $\frac{13}{20}$ ● 0.63

12. 0.5 ● $\frac{3}{5}$

13. DISTANCE River Road is $11\frac{4}{5}$ miles long. Prairie Road is 14.9 miles long. How much longer is Prairie Road than River Road?

14. ANIMALS The table shows lengths of different pond insects. Using decimals, name the insect having the smallest length and the insect having the greatest length.

Pond Insects

Insect	Deer Fly	Spongilla Fly	Springtail	Water Treader
Length (in.)	$\frac{2}{5}$	$\frac{3}{10}$	$\frac{3}{20}$	$\frac{1}{2}$

Source: *Golden Nature Guide to Pond Life*

Copyright © Glencoe/McGraw-Hill, a division of The McGraw-Hill Companies, Inc.

NAME ______________________________ DATE ______________ PERIOD _____

4-9 Study Guide and Intervention

Algebra: Ordered Pairs and Functions

Lesson 4–9

A **coordinate plane** is formed when two number lines intersect at their zero points. This intersection is called the **origin**. The horizontal number line is called the ***x*-axis**. The vertical number line is called the ***y*-axis**.

An **ordered pair** is used to name a point on a coordinate plane. The first number in the ordered pair is the ***x*-coordinate**, and the second number is the ***y*-coordinate**.

Example 1 **Write the ordered pair that names point *A*.**

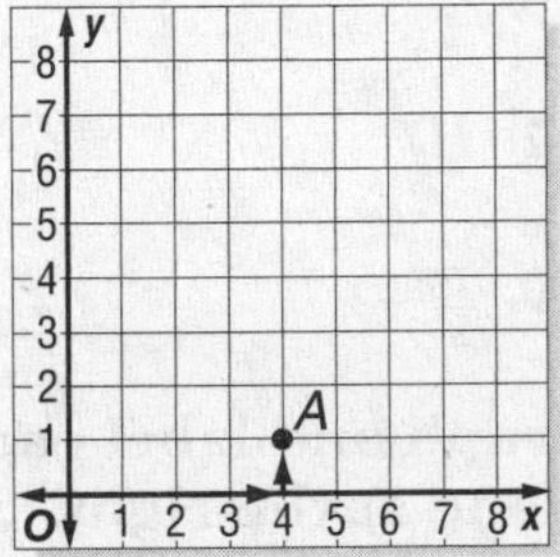

Start at the origin. Move right along the x-axis until you are under point A. The x-coordinate is 4.

Then move up until you reach point A. The y-coordinate is 1.

So, point A is named by the ordered pair (4, 1).

Example 2 **Graph the point *W*(2, 4).**

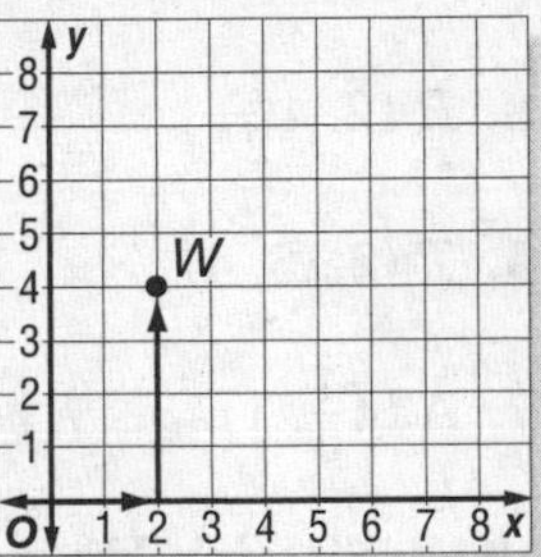

Start at the origin. Move 2 units to the right along the x-axis.

Then move 4 units up to locate the point. Draw a dot and label the point W.

Exercises

Use the coordinate plane at the right to name the ordered pair for each point.

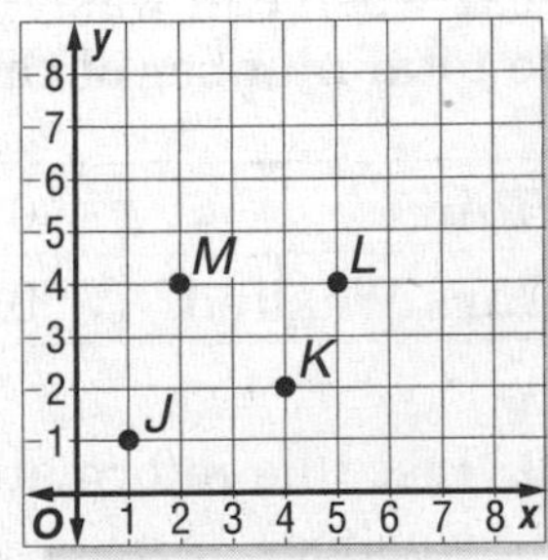

1. J

2. K

3. L

4. M

Graph and label each point on the coordinate plane.

5. $S(1, 3)$

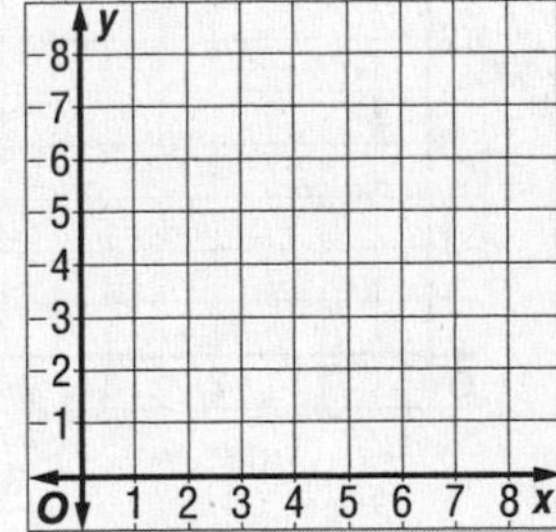

6. $T(4, 0)$

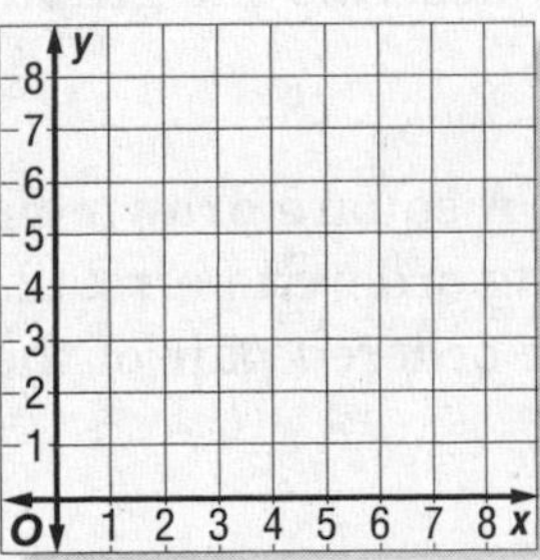

Copyright © Glencoe/McGraw-Hill, a division of The McGraw-Hill Companies, Inc.

4-9 Practice

Algebra: Ordered Pairs and Functions

Use the coordinate plane at the right to name the ordered pair for each point.

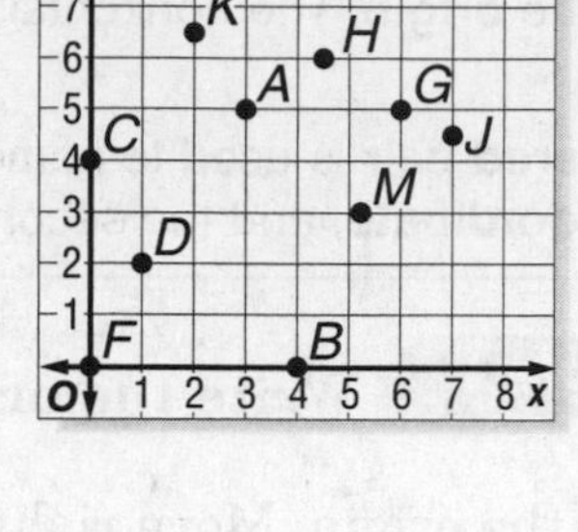

1. *A*

2. *B*

3. *C*

4. *D*

5. *F*

6. *G*

7. *H*

8. *J*

9. *K*

10. *M*

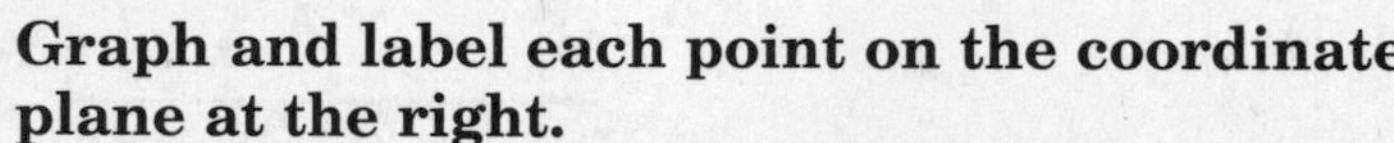

Graph and label each point on the coordinate plane at the right.

11. $N(4, 3)$

12. $P(0, 4)$

13. $R(2, 4\frac{1}{2})$

14. $S(1\frac{3}{4}, 2)$

15. $T(2.75, 4)$

16. $W(3, 1.5)$

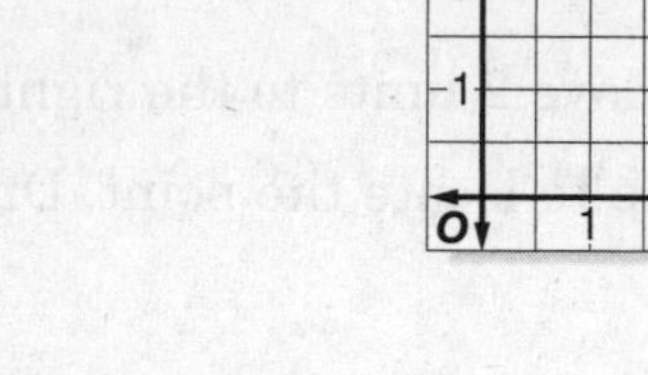

17. $A(4\frac{1}{4}, 1)$

18. $B(1, 3\frac{3}{4})$

CAR WASH For Exercises 19 and 20, use the following information.

A car wash can wash four cars in one hour. The table shows the total number of cars washed in 0, 1, 2, and 3 hours.

Hours	0	1	2	3
Cars Washed	0	4	8	12

19. List this information as ordered pairs (number of hours, number of cars washed).

20. Graph the ordered pairs on the coordinate plane at the right. Then describe the graph.

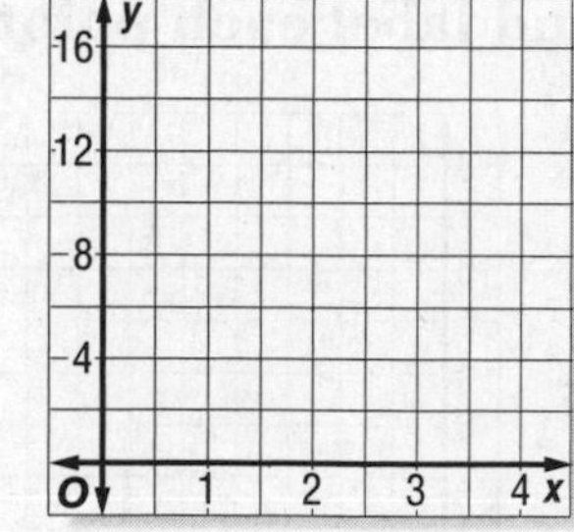

21. GEOMETRY A square drawn on a coordinate plane has the following ordered pairs: (2, 2.5), (2, 6.5), and (6, 2.5). What is the ordered pair of the fourth point?

Copyright © Glencoe/McGraw-Hill, a division of The McGraw-Hill Companies, Inc.

NAME ______________________ DATE ____________ PERIOD _____

5-1 Study Guide and Intervention

Rounding Fractions and Mixed Numbers

Use these guidelines to round fractions and mixed numbers to the nearest half.

	Rounding Fractions and Mixed Numbers	Example
Round up	When the numerator is almost as large as the denominator, round up to the next whole number.	$\frac{7}{8}$ rounds to 1.
Round to $\frac{1}{2}$	When the numerator is about half of the denominator, round the fraction to $\frac{1}{2}$.	$4\frac{3}{7}$ rounds to $4\frac{1}{2}$.
Round down	When the numerator is much smaller than the denominator, round down to the previous whole number.	$\frac{1}{5}$ rounds to 0.

Lesson 5–1

Example 1 **Round $2\frac{3}{8}$ to the nearest half.**

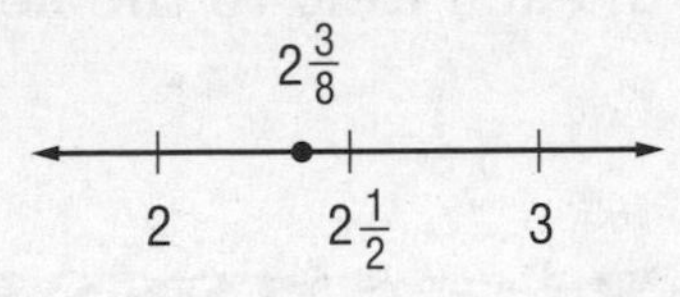

The numerator of $\frac{3}{8}$ is about half of the denominator. So, $2\frac{3}{8}$ rounds to $2\frac{1}{2}$.

Round up when it is better for a measure to be too large than too small. Round down when it is better for a measure to be too small than too large.

Example 2 **MUSIC You have $4\frac{1}{2}$ minutes left on a CD you are making for a friend. Should you choose a song that is 5 minutes long or a song that is 4 minutes long?**

In order for the entire song to be recorded, you should round down the number of minutes left on the CD and choose the song that is 4 minutes long.

Exercises

Round each number to the nearest half.

1. $\frac{2}{5}$ **2.** $\frac{1}{18}$ **3.** $\frac{6}{13}$ **4.** $6\frac{2}{9}$

5. $5\frac{4}{7}$ **6.** $8\frac{4}{5}$ **7.** $4\frac{1}{7}$ **8.** $\frac{2}{7}$

Tell whether each number should be rounded up or down.

9. the weight of a package you are mailing

10. the length of a rug for your bathroom

Copyright © Glencoe/McGraw-Hill, a division of The McGraw-Hill Companies, Inc.

NAME ______________________ DATE __________ PERIOD ______

5-1 Practice

Rounding Fractions and Mixed Numbers

Round each number to the nearest half.

1. $8\frac{1}{7}$ **2.** $\frac{11}{12}$ **3.** $4\frac{3}{8}$ **4.** $2\frac{2}{3}$

5. $6\frac{5}{9}$ **6.** $2\frac{3}{10}$ **7.** $\frac{7}{12}$ **8.** $3\frac{5}{6}$

9. $1\frac{5}{16}$ **10.** $\frac{11}{16}$ **11.** $7\frac{5}{24}$ **12.** $5\frac{25}{32}$

Find the length of each item to the nearest half inch.

13.

14.

15.

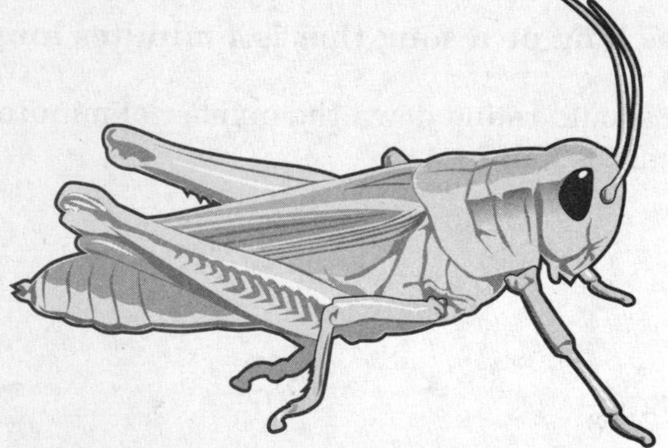

16.

17. STORAGE Mike is moving and is packing his books. His largest book is $12\frac{1}{3}$ inches long. He can choose from two boxes. One box is $12\frac{2}{5}$ inches long and the other box is $12\frac{2}{9}$ inches long. Which size box should Mike use? Why?

18. CRAFTS Gloria is covering the top of a table with colored paper for displaying her pottery. The top of the table is $6\frac{1}{4}$ feet by $3\frac{5}{8}$ feet. To the nearest half foot, how large must the paper be?

Copyright © Glencoe/McGraw-Hill, a division of The McGraw-Hill Companies, Inc.

NAME ______________________ DATE ____________ PERIOD _____

5-2 Study Guide and Intervention

Problem-Solving Investigation: Act It Out

When solving problems, one strategy that is helpful is to *act it out*. By using paper and pencil, a model, fraction strips, or any manipulative, you can often act out the problem situation. Then by using your model, you can determine an answer to the situation.

You can use the *act it out* strategy, along with the following four-step problem-solving plan to solve a problem.

1 Understand – Read and get a general understanding of the problem.

2 Plan – Make a plan to solve the problem and estimate the solution.

3 Solve – Use your plan to solve the problem.

4 Check – Check the reasonableness of your solution.

Example 1 **HOBBIES This fall, Patrick is going to play one sport and take music lessons. He is deciding between playing football, cross country, or soccer. He is also deciding between guitar lessons or piano lessons. How many possible combinations are there of a sport and music lesson for Patrick?**

Understand You know the three sports he is choosing from: football, cross country, and soccer. You also know the music lessons he is choosing from: guitar and piano. You need to determine how many possible combinations there are.

Plan Start by choosing one sport, and pairing it with each of the two music lessons. Then do this for each sport.

Solve

football, guitar	cross country, guitar	soccer, guitar
football, piano	cross country, piano	soccer, piano

So, there are 6 possible combinations of a sport and music lessons.

Check You can multiply the number of sport choices by the number of music lesson choices. $3 \times 2 = 6$

Exercise

WOOD WORK Darnell and his dad are making wooden picture frames. Each picture frame uses $1\frac{1}{4}$ feet of wood. If they have a total of $8\frac{1}{2}$ feet of wood, how many picture frames can they make?

Lesson 5–2

Copyright © Glencoe/McGraw-Hill, a division of The McGraw-Hill Companies, Inc.

NAME ______________________ DATE ____________ PERIOD _____

5-2 Practice

Problem-Solving Investigation: Act It Out

Mixed Problem Solving

Use the act it out strategy to solve Exercises 1 and 2.

1. **FITNESS** Brad jumps 4 feet forward and then 2 feet backward. How many sets will he have jumped when he reaches 16 feet?

2. **SEWING** Dion's grandmother is making a quilt using four small squares put together to form one large square block. How many different blocks can she make using one each of red, green, blue, and yellow small squares? Show the possible arrangements.

Y	B
G	R

R G		G R		B R		Y R	
B Y		B Y		G Y		G B	
R G		G R		B R		Y R	
Y B		Y B		Y G		B G	
R B		G B		B G		Y G	
G Y		R Y		R Y		R B	
n D		Q D		D G		Y C	
Y G		Y R		Y R		B R	
R Y		G Y		B Y		Y B	
G B		R B		R G		R G	
R Y		G Y		B Y			
B G		B R		G R			

Use any strategy to solve Exercises 3–6. Some strategies are shown below.

Problem-Solving Strategies
• Use make a table.
• Use act it out.

3. **ANIMALS** Nine birds are sitting on a power line. Three more birds arrive at the same time five of the birds fly off. How many birds are sitting on the power line now?

4. **MONEY** Ping bought a pair of running shoes for $7 less than the regular price. If he paid $29, what was the regular price?

5. **FOOD** Elena bought three bags of dried fruit that weighed $1\frac{7}{10}$ pounds, $3\frac{1}{4}$ pounds, and $2\frac{3}{5}$ pounds. About how much fruit did she buy?

6. **PATTERNS** What number is missing in the pattern
. . . , 654, 533, □, 291, . . . ?

Copyright © Glencoe/McGraw-Hill, a division of The McGraw-Hill Companies, Inc.

NAME ______________________ DATE ____________ PERIOD ____

5-3 Study Guide and Intervention

Adding and Subtracting Fractions with Like Denominators

Fractions with the same denominator are called **like fractions**.

- To add like fractions, add the numerators. Use the same denominator in the sum.
- To subtract like fractions, subtract the numerators. Use the same denominator in the difference.

Example 1 **Find the sum of $\frac{3}{5}$ and $\frac{3}{5}$.**

Estimate $\frac{1}{2} + \frac{1}{2} = 1$

$\frac{3}{5} + \frac{3}{5} = \frac{3+3}{5}$ Add the numerators.

$= \frac{6}{5}$ Simplify.

$= 1\frac{1}{5}$ Write the improper fraction as a mixed number.

$\frac{3}{5}$ $\frac{3}{5}$ $1\frac{1}{5}$

Compared to the estimate, the answer is reasonable.

Example 2 **Find the difference of $\frac{3}{4}$ and $\frac{1}{4}$.**

Estimate $1 - 0 = 1$

$\frac{3}{4} - \frac{1}{4} = \frac{3-1}{4}$ Subtract the numerators.

$= \frac{2}{4}$ or $\frac{1}{2}$ Simplify.

Compared to the estimate, the answer is reasonable.

Exercises

Add or subtract. Write in simplest form.

1. $\frac{1}{9} + \frac{4}{9}$ **2.** $\frac{9}{11} - \frac{7}{11}$ **3.** $\frac{9}{10} + \frac{5}{10}$ **4.** $\frac{11}{12} - \frac{9}{12}$

5. $\frac{4}{7} + \frac{5}{7}$ **6.** $\frac{4}{9} - \frac{1}{9}$ **7.** $\frac{7}{8} + \frac{5}{8}$ **8.** $\frac{6}{7} - \frac{4}{7}$

9. $\frac{3}{4} + \frac{3}{4}$ **10.** $\frac{4}{5} - \frac{1}{5}$ **11.** $\frac{5}{6} + \frac{1}{6}$ **12.** $\frac{7}{10} - \frac{1}{10}$

Lesson 5–3

Copyright © Glencoe/McGraw-Hill, a division of The McGraw-Hill Companies, Inc.

NAME ______________________ DATE ____________ PERIOD ______

5-3 Practice

Adding and Subtracting Fractions with Like Denominators

Add or subtract. Write in simplest form.

1. $\frac{3}{7} + \frac{6}{7}$ **2.** $\frac{2}{5} + \frac{4}{5}$ **3.** $\frac{3}{4} + \frac{3}{4}$ **4.** $\frac{2}{3} + \frac{2}{3}$

5. $\frac{5}{8} + \frac{7}{8}$ **6.** $\frac{11}{16} + \frac{7}{16}$ **7.** $\frac{7}{8} - \frac{3}{8}$ **8.** $\frac{3}{10} - \frac{1}{10}$

9. $\frac{11}{15} - \frac{6}{15}$ **10.** $\frac{7}{9} - \frac{4}{9}$ **11.** $\frac{9}{11} - \frac{6}{11}$ **12.** $\frac{17}{18} - \frac{5}{18}$

13. $\frac{5}{7} + \frac{1}{7} + \frac{6}{7}$ **14.** $\frac{9}{10} + \frac{9}{10} - \frac{3}{10}$ **15.** $\frac{11}{12} - \frac{7}{12} + \frac{5}{12}$

Write an addition or subtraction expression for each model. Then add or subtract.

16.

17.

18. **WEATHER** In January through March, Death Valley gets a total of about $\frac{21}{25}$ inch of precipitation. In April through June, it gets a total of about $\frac{6}{25}$ inch. How much more precipitation occurs in January through March?

19. **ANALYZE GRAPHS** What part of the school population likes basketball, baseball, or football? How much larger is this than the part of the student population that prefers soccer?

Brown Middle School's Favorite Sports

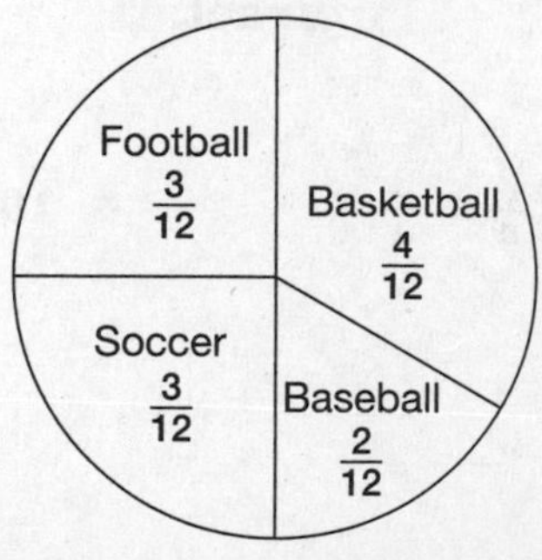

Copyright © Glencoe/McGraw-Hill, a division of The McGraw-Hill Companies, Inc.

NAME ______________________ DATE ____________ PERIOD _____

5-4 Study Guide and Intervention

Adding and Subtracting Fractions with Unlike Denominators

To find the sum or difference of two fractions with unlike denominators, rename the fractions using the least common denominator (LCD). Then add or subtract and simplify.

Example 1 **Find $\frac{1}{3} + \frac{5}{6}$.**

The LCD of $\frac{1}{3}$ and $\frac{5}{6}$ is 6.

Write the problem.		Rename using the LCD, 6.		Add the fractions.
$\frac{1}{3}$ $+\frac{5}{6}$	→	$\frac{1 \times 2}{3 \times 2} = \frac{2}{6}$ $+\frac{5 \times 1}{6 \times 1} = \frac{5}{6}$	→	$\frac{2}{6}$ $+\frac{5}{6}$ $\frac{7}{6}$ or $1\frac{1}{6}$

Example 2 **Find $\frac{2}{3} - \frac{1}{4}$.**

The LCD of $\frac{2}{3}$ and $\frac{1}{4}$ is 12.

Write the problem.		Rename using the LCD, 12.		Subtract the fractions.
$\frac{2}{3}$ $-\frac{1}{4}$	→	$\frac{2 \times 4}{3 \times 4} = \frac{8}{12}$ $-\frac{1 \times 3}{4 \times 3} = \frac{3}{12}$	→	$\frac{8}{12}$ $-\frac{3}{12}$ $\frac{5}{12}$

Example 3 **Evaluate $x - y$ if $x = \frac{1}{2}$ and $y = \frac{2}{5}$.**

$x - y = \frac{1}{2} - \frac{2}{5}$ — Replace x with $\frac{1}{2}$ and y with $\frac{2}{5}$.

$= \frac{1 \times 5}{2 \times 5} - \frac{2 \times 2}{5 \times 2}$ — Rename $\frac{1}{2}$ and $\frac{2}{5}$ using the LCD, 10.

$= \frac{5}{10} - \frac{4}{10}$ — Simplify.

$= \frac{1}{10}$ — Subtract the numerators.

Exercises

Add or subtract. Write in simplest form.

1. $\frac{1}{6} + \frac{1}{2}$ **2.** $\frac{2}{3} - \frac{1}{2}$ **3.** $\frac{1}{4} + \frac{7}{8}$ **4.** $\frac{9}{10} - \frac{3}{5}$

5. $\frac{2}{7} + \frac{1}{2}$ **6.** $\frac{5}{6} - \frac{1}{12}$ **7.** $\frac{7}{10} + \frac{1}{2}$ **8.** $\frac{4}{9} - \frac{1}{3}$

9. Evaluate $x + y$ if $x = \frac{1}{12}$ and $y = \frac{1}{6}$.

10. Evaluate $a + b$ if $a = \frac{1}{2}$ and $b = \frac{3}{4}$.

Lesson 5–4

Copyright © Glencoe/McGraw-Hill, a division of The McGraw-Hill Companies, Inc.

NAME ______________________ DATE __________ PERIOD ____

5-4 Practice

Adding and Subtracting Fractions with Unlike Denominators

Add or subtract. Write in simplest form.

1. $\frac{3}{4} + \frac{1}{8}$

2. $\frac{1}{2} + \frac{1}{3}$

3. $\frac{11}{12} - \frac{2}{3}$

4. $\frac{7}{10} - \frac{1}{2}$

5. $\frac{1}{6} + \frac{3}{10}$

6. $\frac{3}{4} + \frac{1}{6}$

7. $\frac{3}{5} - \frac{1}{4}$

8. $\frac{6}{7} - \frac{3}{4}$

9. $\frac{6}{7} + \frac{1}{3}$

10. $\frac{9}{10} + \frac{3}{5}$

11. $\frac{11}{12} - \frac{3}{4}$

12. $\frac{7}{11} - \frac{1}{2}$

ALGEBRA Evaluate each expression.

13. $a + b$ if $a = \frac{3}{5}$ and $b = \frac{5}{8}$

14. $c - d$ if $c = \frac{9}{10}$ and $d = \frac{5}{6}$

15. ANIMALS A newborn panda at the San Diego zoo grew about $\frac{9}{16}$ pound the first week and about $\frac{5}{8}$ pound the second week. How much more did the panda grow the second week? Justify your answer.

16. EXERCISES Every day Kim does leg muscle exercises for $\frac{3}{7}$ of an hour and foot muscle exercises for $\frac{2}{3}$ of an hour. Which exercises does she spend the most time doing and by how much?

Copyright © Glencoe/McGraw-Hill, a division of The McGraw-Hill Companies, Inc.

5-5 Study Guide and Intervention

Adding and Subtracting Mixed Numbers

To add or subtract mixed numbers:

1. Add or subtract the fractions.
2. Then add or subtract the whole numbers.
3. Rename and simplify if necessary.

Example 1 **Find $2\frac{1}{3} + 4\frac{1}{4}$.**

The LCD of $\frac{1}{3}$ and $\frac{1}{4}$ is 12.

Write the problem.

$$\begin{array}{r} 2\frac{1}{3} \\ +\ 4\frac{1}{4} \\ \hline \end{array}$$

→ Rename the fractions using the LCD, 12.

$$\begin{array}{rcr} 2\frac{1 \times 4}{3 \times 4} & = & 2\frac{4}{12} \\ +\ 4\frac{1 \times 3}{4 \times 3} & = & +\ 4\frac{3}{12} \\ \hline \end{array}$$

→ Add the fractions. Then add the whole numbers.

$$\begin{array}{r} 2\frac{4}{12} \\ +\ 4\frac{3}{12} \\ \hline 6\frac{7}{12} \end{array}$$

So, $2\frac{1}{3} + 4\frac{1}{4} = 6\frac{7}{12}$.

Example 2 **Find $6\frac{1}{2} - 2\frac{1}{3}$.**

The LCD of $\frac{1}{2}$ and $\frac{1}{3}$ is 6.

Write the problem.

$$\begin{array}{r} 6\frac{1}{2} \\ -\ 2\frac{1}{3} \\ \hline \end{array}$$

→ Rename the fractions using the LCD, 6.

$$\begin{array}{rcr} 6\frac{1 \times 3}{2 \times 3} & = & 6\frac{3}{6} \\ -\ 2\frac{1 \times 2}{3 \times 2} & = & -\ 2\frac{2}{6} \\ \hline \end{array}$$

→ Subtract the fractions. Then subtract the whole numbers.

$$\begin{array}{r} 6\frac{3}{6} \\ -\ 2\frac{2}{6} \\ \hline 4\frac{1}{6} \end{array}$$

So, $6\frac{1}{2} - 2\frac{1}{3} = 4\frac{1}{6}$.

Exercises

Add or subtract. Write in simplest form.

1. $\begin{array}{r} 3\frac{2}{3} \\ -\ 2\frac{1}{3} \\ \hline \end{array}$

2. $\begin{array}{r} 4\frac{3}{4} \\ +\ 1\frac{3}{4} \\ \hline \end{array}$

3. $\begin{array}{r} 5\frac{1}{2} \\ +\ 4\frac{1}{3} \\ \hline \end{array}$

4. $\begin{array}{r} 6\frac{7}{8} \\ -\ 3\frac{1}{2} \\ \hline \end{array}$

5. $3\frac{2}{3} - 1\frac{1}{2}$

6. $4\frac{2}{3} + 2\frac{1}{4}$

7. $5\frac{1}{3} - 2\frac{1}{4}$

Copyright © Glencoe/McGraw-Hill, a division of The McGraw-Hill Companies, Inc.

NAME ______________________________ DATE ____________ PERIOD _____

5-5 Practice

Adding and Subtracting Mixed Numbers

Add or subtract. Write in simplest form.

1. $5 - 3\frac{4}{7}$

2. $8 - 2\frac{3}{8}$

3. $7\frac{7}{8} - 3\frac{3}{8}$

4. $8\frac{5}{7} - 4\frac{3}{7}$

5. $9\frac{3}{4} - 2\frac{3}{8}$

6. $6\frac{2}{3} - 1\frac{1}{6}$

7. $8\frac{1}{4} + 2\frac{4}{5}$

8. $10\frac{2}{3} + 8\frac{7}{10}$

9. $5\frac{9}{10} + 3\frac{1}{2}$

10. $3\frac{5}{6} + 10\frac{5}{8}$

11. $8\frac{5}{6} - 3\frac{1}{3}$

12. $9\frac{6}{7} - 2\frac{5}{14}$

ALGEBRA. Evaluate each expression if $a = 3\frac{5}{6}$, $b = 2\frac{2}{3}$, *and* $c = 1\frac{1}{4}$.

13. $a + b$

14. $a + c$

15. $b - c$

16. $a - c$

17. COOKING A punch recipe calls for $4\frac{1}{4}$ cups pineapple juice, $2\frac{2}{3}$ cups orange juice, and $3\frac{1}{2}$ cups cranberry juice. How much juice is needed to make the punch?

18. ANALYZE TABLES The wingspans of two butterflies and a moth are shown. How much greater is the longest wingspan than the shortest wingspan? Justify your answer.

Butterfly and Moth Wingspans	
Butterfly or Moth	**Width (in.)**
American Snout butterfly	$1\frac{3}{8}$
Garden Tiger moth	$1\frac{13}{16}$
Milbert's Tortoiseshell butterfly	$1\frac{3}{4}$

Copyright © Glencoe/McGraw-Hill, a division of The McGraw-Hill Companies, Inc.

NAME ______________________ DATE ____________ PERIOD _____

5-6 Study Guide and Intervention

Estimating Products of Fractions

Lesson 5–6

Numbers that are easy to divide mentally are called **compatible numbers**. One way to estimate products involving fractions is to use compatible numbers.

Example 1 **Estimate $\frac{2}{3} \times 8$.**

Estimate $\frac{2}{3} \times 8$. Make it easier by finding $\frac{1}{3} \times 8$ first.

$\frac{1}{3} \times 9 = ?$ Change 8 to 9 since 3 and 9 are compatible numbers.

$\frac{1}{3} \times 9 = 3$ $\frac{1}{3}$ of 9, or 9 divided by 3, is 3.

$\frac{2}{3} \times 9 = 6$ Since $\frac{1}{3}$ of 9 is 3, $\frac{2}{3}$ of 9 is 2×3 or 6.

So, $\frac{2}{3} \times 8$ is about 6.

You can estimate the product of fractions by rounding to 0, $\frac{1}{2}$, or 1.

Example 2 **Estimate $\frac{1}{3} \times \frac{5}{6}$.**

$\frac{1}{3} \times \frac{5}{6} \rightarrow \frac{1}{2} \times 1 = \frac{1}{2}$.

So, $\frac{1}{3} \times \frac{5}{6}$ is about $\frac{1}{2}$.

You can estimate the product of mixed numbers by rounding to the next whole number.

Example 3 **Estimate $3\frac{1}{4} \times 5\frac{7}{8}$.**

Since $3\frac{1}{4}$ rounds to 3 and $5\frac{7}{8}$ rounds to 6, $3\frac{1}{4} \times 5\frac{7}{8} \rightarrow 3 \times 6 = 18$.

So, $3\frac{1}{4} \times 5\frac{7}{8}$ is about 18.

Exercises

Estimate each product. Show how you found your estimate.

1. $\frac{1}{5} \times 24$ **2.** $\frac{7}{8} \times \frac{3}{5}$ **3.** $7\frac{2}{7} \times 5\frac{3}{4}$ **4.** $\frac{4}{7} \times 20$

5. $\frac{5}{8} \times 19$ **6.** $2\frac{4}{5} \times 6\frac{1}{12}$ **7.** $\frac{1}{9} \times \frac{1}{12}$ **8.** $3\frac{7}{8} \times 10\frac{1}{10}$

9. $\frac{11}{12} \times \frac{6}{7}$ **10.** $\frac{3}{8} \times 17$ **11.** $4\frac{7}{8} \times 2\frac{9}{10}$ **12.** $\frac{11}{12} \times \frac{1}{3}$

Copyright © Glencoe/McGraw-Hill, a division of The McGraw-Hill Companies, Inc.

NAME ______________________ DATE ____________ PERIOD _____

5-6 Practice

Estimating Products of Fractions

Estimate each product.

1. $\frac{1}{3} \times 28$

2. $\frac{1}{7} \times 20$

3. $\frac{1}{9}$ of 83

4. $\frac{1}{6}$ of 23

5. $\frac{2}{3} \times 76$

6. $\frac{3}{8} \times 15$

7. $\frac{2}{5}$ of 37

8. $\frac{2}{3}$ of 11

9. $\frac{3}{5} \times \frac{2}{9}$

10. $\frac{7}{8} \times \frac{4}{5}$

11. $\frac{10}{19} \times \frac{3}{8}$

12. $\frac{3}{4} \times \frac{3}{7}$

13. $\frac{6}{7} \times \frac{1}{4}$

14. $2\frac{9}{10} \times 6\frac{1}{4}$

15. $4\frac{3}{8} \times 7\frac{2}{7}$

Estimate the area of each rectangle.

16. Rectangle: $2\frac{1}{3}$ m by $6\frac{5}{8}$ m

17. Rectangle: $4\frac{7}{9}$ in. by $5\frac{2}{7}$ in.

SCULPTURE Trevor is using the recipe for a sculpture carving material shown at the right.

Girostone Recipe
1 cup vermiculite
$\frac{1}{4}$ cup cement
$\frac{1}{8}$ cup sand
water to form thick paste

18. About how many cups of cement would he need to make $\frac{4}{9}$ batch of the recipe?

19. About how many cups of sand would he need to make $1\frac{6}{7}$ batches of the recipe?

Copyright © Glencoe/McGraw-Hill, a division of The McGraw-Hill Companies, Inc.

NAME ______________________ DATE ____________ PERIOD _____

5-7 Study Guide and Intervention

Multiplying Fractions

Type of Product	What To Do	Example
two fractions	Multiply the numerators. Then multiply the denominators.	$\frac{2}{3} \times \frac{4}{5} = \frac{2 \times 4}{3 \times 5} = \frac{8}{15}$
fraction and a whole number	Rename the whole number as an improper fraction. Multiply the numerators. Then multiply the denominators.	$\frac{3}{11} \times 6 = \frac{3}{11} \times \frac{6}{1} = \frac{18}{11} = 1\frac{7}{11}$

Lesson 5–7

Example 1 **Find $\frac{2}{5} \times \frac{3}{4}$.** **Estimate:** $\frac{1}{2} \times 1 = \frac{1}{2}$

$\frac{2}{5} \times \frac{3}{4} = \frac{2 \times 3}{5 \times 4}$ Multiply the numerators. Multiply the denominators.

$= \frac{6}{20}$ or $\frac{3}{10}$ Simplify. Compare to the estimate.

Example 2 **Find $\frac{4}{9} \times 8$.** **Estimate:** $\frac{1}{2} \times 8 = 4$

$\frac{4}{9} \times 8 = \frac{4}{9} \times \frac{8}{1}$ Write 8 as $\frac{8}{1}$.

$= \frac{4 \times 8}{9 \times 1}$ Multiply.

$= \frac{32}{9}$ or $3\frac{5}{9}$ Simplify. Compare to the estimate.

Example 3 **Find $\frac{2}{5} \times \frac{3}{8}$.** **Estimate:** $\frac{1}{2} \times \frac{1}{2} = \frac{1}{4}$

$\frac{2}{5} \times \frac{3}{8} = \frac{\overset{1}{\cancel{2}} \times 3}{5 \times \underset{4}{\cancel{8}}}$ Divide both the numerator and denominator by the common factor, 2.

$= \frac{3}{20}$ Simplify. Compare to the estimate.

Exercises

Multiply.

1. $\frac{1}{4} \times \frac{5}{6}$ **2.** $\frac{3}{7} \times \frac{3}{4}$ **3.** $4 \times \frac{1}{5}$ **4.** $\frac{5}{12} \times 2$

5. $\frac{3}{5} \times 10$ **6.** $\frac{2}{3} \times \frac{3}{8}$ **7.** $\frac{1}{7} \times \frac{1}{7}$ **8.** $\frac{2}{9} \times \frac{1}{2}$

Copyright © Glencoe/McGraw-Hill, a division of The McGraw-Hill Companies, Inc.

NAME ______________________ DATE ____________ PERIOD ______

5-7 Practice

Multiplying Fractions

Multiply.

1. $\frac{1}{4} \times \frac{3}{5}$

2. $\frac{7}{8} \times \frac{1}{3}$

3. $\frac{1}{2} \times \frac{3}{4}$

4. $\frac{2}{3} \times \frac{2}{9}$

5. $\frac{1}{3} \times 11$

6. $\frac{1}{2} \times 12$

7. $\frac{5}{6} \times 21$

8. $\frac{3}{4} \times 10$

9. $\frac{1}{4} \times \frac{4}{5}$

10. $\frac{4}{9} \times \frac{3}{8}$

11. $\frac{7}{10} \times \frac{4}{21}$

12. $\frac{3}{5} \times \frac{5}{12}$

13. $\frac{1}{3} \times \frac{1}{4} \times \frac{1}{5}$

14. $\frac{3}{4} \times \frac{3}{8} \times \frac{2}{3}$

15. $\frac{2}{3} \times \frac{12}{17} \times \frac{1}{4}$

ALGEBRA Evaluate each expression if $a = \frac{4}{5}$, $b = \frac{1}{2}$, and $c = \frac{2}{7}$.

16. bc

17. abc

18. $ab + \frac{3}{5}$

19. PRESIDENTS By 2005, 42 different men had been President of the United States. Of these men, $\frac{2}{21}$ had no children. How many presidents had no children?

Copyright © Glencoe/McGraw-Hill, a division of The McGraw-Hill Companies, Inc.

NAME ______________________ DATE ____________ PERIOD ______

5-8 Study Guide and Intervention

Multiplying Mixed Numbers

To multiply mixed numbers, write the mixed numbers as improper fractions, and then multiply as with fractions.

Example 1 **Find $2\frac{1}{4} \times 1\frac{2}{3}$.** **Estimate:** 2 × 2 = 4.

$2\frac{1}{4} \times 1\frac{2}{3} = \frac{9}{4} \times \frac{5}{3}$ Write mixed numbers as improper fractions.

$= \frac{\overset{3}{\cancel{9}} \times 5}{4 \times \underset{1}{\cancel{3}}}$ Divide the numerator and denominator by their common factor, 3.

$= \frac{15}{4}$ or $3\frac{3}{4}$ Simplify. Compare to the estimate.

Example 2 **If $a = 1\frac{1}{3}$ and $b = 2\frac{1}{4}$, what is the value of ab?**

$ab = 1\frac{1}{3} \times 2\frac{1}{4}$ Replace *a* with $1\frac{1}{3}$ and *b* with $2\frac{1}{4}$.

$= \frac{4}{3} \times \frac{9}{4}$ Write mixed numbers as improper fractions.

$= \frac{\overset{1}{\cancel{4}}}{\underset{1}{\cancel{3}}} \times \frac{\overset{3}{\cancel{9}}}{\underset{1}{\cancel{4}}}$ Divide the numerator and denominator by their common factors, 3 and 4.

$= \frac{3}{1}$ or 3 Simplify.

Exercises

Multiply. Write in simplest form.

1. $\frac{1}{3} \times 1\frac{1}{3}$ **2.** $1\frac{1}{5} \times \frac{3}{4}$ **3.** $3 \times 1\frac{3}{5}$ **4.** $\frac{2}{3} \times 3\frac{1}{2}$

5. $9 \times 1\frac{1}{6}$ **6.** $2\frac{4}{9} \times \frac{4}{11}$ **7.** $2\frac{1}{2} \times 1\frac{1}{3}$ **8.** $1\frac{1}{4} \times \frac{3}{5}$

9. $8 \times 1\frac{1}{4}$ **10.** $\frac{3}{8} \times 2\frac{1}{2}$ **11.** $4 \times 1\frac{1}{8}$ **12.** $1\frac{1}{9} \times 3$

13. ALGEBRA Evaluate $5x$ if $x = 1\frac{2}{3}$.

14. ALGEBRA If $t = 2\frac{3}{8}$, what is $4t$?

Lesson 5–8

Copyright © Glencoe/McGraw-Hill, a division of The McGraw-Hill Companies, Inc.

NAME ______________________________ DATE ____________ PERIOD _____

5-8 Practice

Multiplying Mixed Numbers

Multiply. Write in simplest form.

1. $\frac{4}{5} \times 3\frac{1}{8}$

2. $\frac{9}{10} \times 3\frac{1}{3}$

3. $1\frac{3}{5} \times \frac{3}{5}$

4. $2\frac{5}{8} \times \frac{2}{3}$

5. $\frac{2}{3} \times 3\frac{1}{4}$

6. $3\frac{3}{4} \times 2\frac{2}{3}$

7. $1\frac{1}{4} \times 2\frac{2}{3}$

8. $5\frac{1}{3} \times 2\frac{1}{4}$

9. $2\frac{1}{5} \times 1\frac{1}{4}$

10. $5\frac{1}{2} \times 4\frac{1}{3}$

11. $\frac{2}{9} \times \frac{3}{4} \times 2\frac{1}{4}$

12. $1\frac{1}{2} \times 2\frac{1}{6} \times 1\frac{1}{5}$

ALGEBRA **Evaluate each expression if $f = \frac{6}{7}$, $g = 1\frac{3}{4}$, and $h = 2\frac{2}{3}$.**

13. fg

14. $\frac{3}{8}h$

15. gh

16. LUMBER A lumber yard has a scrap sheet of plywood that is $23\frac{3}{4}$ inches by $41\frac{1}{5}$ inches. What is the area of the plywood?

17. LANDSCAPING A planter box in the city plaza measures $3\frac{2}{3}$ feet by $4\frac{1}{8}$ feet by $2\frac{1}{2}$ feet. Find the volume of the planter box.

Copyright © Glencoe/McGraw-Hill, a division of The McGraw-Hill Companies, Inc.

NAME ______________________________ DATE ____________ PERIOD _____

5-9 Study Guide and Intervention

Dividing Fractions

When the product of two numbers is 1, the numbers are called **reciprocals**.

Example 1 **Find the reciprocal of 8.**

Since $8 \times \frac{1}{8} = 1$, the reciprocal of 8 is $\frac{1}{8}$.

Example 2 **Find the reciprocal of $\frac{5}{9}$.**

Since $\frac{5}{9} \times \frac{9}{5} = 1$, the reciprocal of $\frac{5}{9}$ is $\frac{9}{5}$.

You can use reciprocals to divide fractions. To divide by a fraction, multiply by its reciprocal.

Example 3 **Find $\frac{2}{3} \div \frac{4}{5}$.**

$\frac{2}{3} \div \frac{4}{5} = \frac{2}{3} \times \frac{5}{4}$ — Multiply by the reciprocal, $\frac{5}{4}$.

$= \frac{\overset{1}{\cancel{2}} \times 5}{3 \times \underset{2}{\cancel{4}}}$ — Divide 2 and 4 by the GCF, 2.

$= \frac{5}{6}$ — Multiply numerators and denominators.

Exercises

Find the reciprocal of each number.

1. 2 **2.** $\frac{1}{6}$ **3.** $\frac{4}{11}$ **4.** $\frac{3}{5}$

Divide. Write in simplest form.

5. $\frac{1}{3} \div \frac{2}{5}$ **6.** $\frac{1}{9} \div \frac{1}{2}$ **7.** $\frac{2}{3} \div \frac{1}{4}$ **8.** $\frac{1}{2} \div \frac{3}{4}$

9. $\frac{4}{5} \div 2$ **10.** $\frac{4}{5} \div \frac{1}{10}$ **11.** $\frac{5}{12} \div \frac{5}{6}$ **12.** $\frac{9}{10} \div 3$

13. $\frac{3}{4} \div \frac{7}{12}$ **14.** $\frac{9}{10} \div 9$ **15.** $\frac{2}{3} \div \frac{5}{8}$ **16.** $4 \div \frac{7}{9}$

Lesson 5–9

Copyright © Glencoe/McGraw-Hill, a division of The McGraw-Hill Companies, Inc.

NAME ______________________________ DATE ____________ PERIOD _____

5-9 Practice

Dividing Fractions

Find the reciprocal of each number.

1. $\frac{2}{7}$ **2.** $\frac{1}{9}$ **3.** $\frac{3}{8}$ **4.** 2 **5.** 12

Divide. Write in simplest form.

6. $\frac{2}{3} \div \frac{1}{6}$ **7.** $\frac{1}{2} \div \frac{2}{5}$ **8.** $\frac{2}{3} \div \frac{1}{4}$

9. $\frac{3}{4} \div \frac{1}{10}$ **10.** $2 \div \frac{1}{4}$ **11.** $8 \div \frac{2}{5}$

12. $3 \div \frac{4}{5}$ **13.** $2 \div \frac{5}{8}$ **14.** $\frac{3}{7} \div 3$

15. $\frac{4}{5} \div 10$ **16.** $\frac{7}{9} \div 14$ **17.** $\frac{5}{7} \div 4$

ALGEBRA **Find the value of each expression if $h = \frac{3}{8}$, $j = \frac{1}{3}$, and $k = \frac{1}{4}$.**

18. $h \div k$ **19.** $k \div j - h$ **20.** $h \div j + k$

21. INSECTS An average ant is $\frac{1}{4}$ inch long. An average aphid is $\frac{3}{32}$ inch long. How many times longer is an average ant than an average aphid?

Copyright © Glencoe/McGraw-Hill, a division of The McGraw-Hill Companies, Inc.

NAME ______________________ DATE ____________ PERIOD _____

5-10 Study Guide and Intervention

Dividing Mixed Numbers

To divide mixed numbers, express each mixed number as an improper fraction. Then divide as with fractions.

Example 1 **Find $2\frac{2}{3} \div 1\frac{1}{5}$.** **Estimate:** $3 \div 1 = 3$

$2\frac{2}{3} \div 1\frac{1}{5} = \frac{8}{3} \div \frac{6}{5}$ Write mixed numbers as improper fractions.

$= \frac{8}{3} \times \frac{5}{6}$ Multiply by the reciprocal, $\frac{5}{6}$.

$= \frac{\overset{4}{\cancel{8}} \times 5}{3 \times \underset{3}{\cancel{6}}}$ Divide 8 and 6 by the GCF, 2.

$= \frac{20}{9}$ or $2\frac{2}{9}$ Simplify. Compare to the estimate.

Example 2 **Find the value of $s \div t$ if $s = 1\frac{2}{3}$ and $t = \frac{3}{4}$.**

$s \div t = 1\frac{2}{3} \div \frac{3}{4}$ Replace s with $1\frac{2}{3}$ and t with $\frac{3}{4}$.

$= \frac{5}{3} \div \frac{3}{4}$ Write $1\frac{2}{3}$ as an improper fraction.

$= \frac{5}{3} \times \frac{4}{3}$ Multiply by the reciprocal, $\frac{4}{3}$.

$= \frac{20}{9}$ or $2\frac{2}{9}$ Simplify.

Exercises

Divide. Write in simplest form.

1. $2\frac{1}{2} \div \frac{4}{5}$ **2.** $1\frac{2}{3} \div 1\frac{1}{4}$ **3.** $5 \div 1\frac{3}{7}$ **4.** $2\frac{1}{3} \div \frac{7}{9}$

5. $5\frac{2}{5} \div \frac{9}{10}$ **6.** $7\frac{1}{2} \div 1\frac{2}{3}$ **7.** $3\frac{5}{6} \div 2$ **8.** $2\frac{1}{4} \div \frac{2}{7}$

9. $9 \div 1\frac{1}{9}$ **10.** $\frac{4}{5} \div 2\frac{6}{7}$ **11.** $1\frac{8}{9} \div 5$ **12.** $\frac{3}{8} \div 2\frac{1}{4}$

13. ALGEBRA If $x = 1\frac{1}{4}$ and $y = 3$, what is $x \div y$?

14. ALGEBRA Evaluate $18 \div t$ if $t = \frac{9}{11}$.

Lesson 5–10

Copyright © Glencoe/McGraw-Hill, a division of The McGraw-Hill Companies, Inc.

NAME ______________________ DATE __________ PERIOD _____

5-10 Practice

Dividing Mixed Numbers

Divide. Write in simplest form.

1. $3\frac{2}{3} \div 2$ **2.** $10 \div 1\frac{1}{4}$ **3.** $4\frac{3}{4} \div \frac{7}{8}$ **4.** $1\frac{15}{16} \div \frac{7}{8}$

5. $7\frac{1}{2} \div 1\frac{1}{4}$ **6.** $3\frac{3}{8} \div 2\frac{1}{4}$ **7.** $2\frac{1}{10} \div 1\frac{1}{5}$ **8.** $4\frac{1}{2} \div 2\frac{7}{10}$

ALGEBRA Evaluate the expression if $r = 2\frac{4}{5}$, $s = 1\frac{3}{4}$, and $t = \frac{2}{3}$.

9. $t \div 10$ **10.** $s \div t$ **11.** $r \div s$ **12.** $r \div (st)$

13. PIPES How many $\frac{3}{4}$-foot lengths of pipe can be cut from a $6\frac{1}{3}$-foot pipe?

14. TRUCKING A truck driver drove 300 miles in $6\frac{3}{4}$ hours. How many miles per hour did the driver drive?

Copyright © Glencoe/McGraw-Hill, a division of The McGraw-Hill Companies, Inc.

NAME ______________________ DATE ____________ PERIOD _____

6-1 Study Guide and Intervention

Ratios and Rates

A **ratio** is a comparison of two numbers by division. A common way to express a ratio is as a fraction in simplest form. Ratios can also be written in other ways. For example, the ratio $\frac{2}{3}$ can be written as 2 to 3, 2 out of 3, or 2:3.

Examples **Refer to the diagram at the right.**

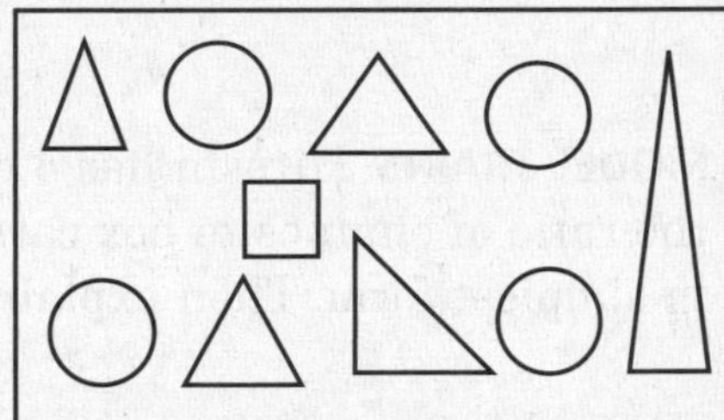

1 Write the ratio in simplest form that compares the number of circles to the number of triangles.

$\begin{matrix}\text{circles} \rightarrow \\ \text{triangles} \rightarrow\end{matrix} \frac{4}{5}$ The GCF of 4 and 5 is 1.

So, the ratio of circles to triangles is $\frac{4}{5}$, 4 to 5, or 4:5.
For every 4 circles, there are 5 triangles.

2 Write the ratio in simplest form that compares the number of circles to the total number of figures.

$\begin{matrix}\text{circles} \rightarrow \\ \text{total figures} \rightarrow\end{matrix} \frac{4}{10} \overset{\div 2}{=} \frac{2}{5}$ The GCF of 4 and 10 is 2.

The ratio of circles to the total number of figures is $\frac{2}{5}$, 2 to 5, or 2:5.
For every two circles, there are five total figures.

A **rate** is a ratio of two measurements having different kinds of units. When a rate is simplified so that it has a denominator of 1, it is called a **unit rate**.

Example 3 **Write the ratio *20 students to 5 computers* as a unit rate.**

$\frac{20 \text{ students}}{5 \text{ computers}} \overset{\div 5}{=} \frac{4 \text{ students}}{1 \text{ computer}}$ Divide the numerator and the denominator by 5 to get a denominator of 1.

The ratio written as a unit rate is *4 students to 1 computer*.

Exercises

Write each ratio as a fraction in simplest form.

1. 2 guppies out of 6 fish **2.** 12 puppies to 15 kittens **3.** 5 boys out of 10 students

Write each rate as a unit rate.

4. 6 eggs for 3 people **5.** $12 for 4 pounds **6.** 40 pages in 8 days

Copyright © Glencoe/McGraw-Hill, a division of The McGraw-Hill Companies, Inc.

NAME ________________________________ DATE ____________ PERIOD _____

6-1 Practice

Ratios and Rates

1. **FRUITS** Find the ratio of bananas to oranges in the graphic at the right. Write the ratio as a fraction in simplest form. Then explain its meaning.

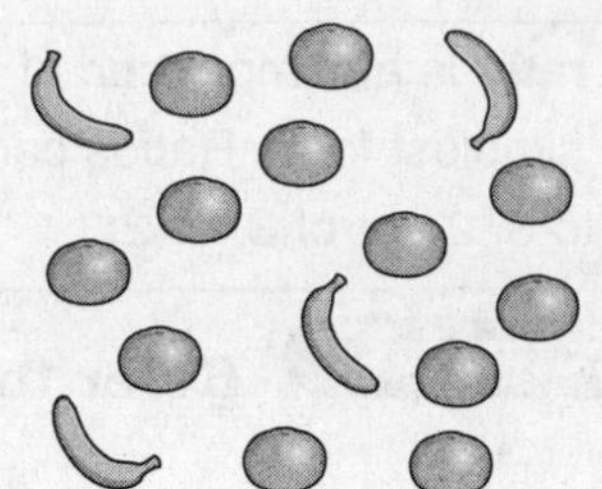

2. **MODEL TRAINS** Hiroshi has 4 engines and 18 box cars. Find the ratio of engines to box cars. Write the ratio as a fraction in simplest form. Then explain its meaning.

3. **ZOOS** A petting zoo has 5 lambs, 11 rabbits, 4 goats, and 4 piglets. Find the ratio of goats to the total number of animals. Then explain its meaning.

4. **FOOD** At the potluck, there were 6 pecan pies, 7 lemon pies, 13 cherry pies, and 8 apple pies. Find the ratio of apple pies to the total number of pies. Then explain its meaning.

Write each rate as a unit rate.

5. 3 inches of snow in 6 hours

6. $46 for 5 toys

7. **TRAINS** The Nozomi train in Japan can travel 558 miles in 3 hours. At this rate, how far can the train travel per hour?

ANALYZE TABLES **For Exercises 8 and 9, refer to the table showing tide pool animals.**

Animals Found in a Tide Pool	
Animal	**Number**
Anemones	11
Limpets	14
Snails	18
Starfish	9

8. Find the ratio of limpets to snails. Then explain its meaning.

9. Find the ratio of snails to the total number of animals. Then explain its meaning.

Copyright © Glencoe/McGraw-Hill, a division of The McGraw-Hill Companies, Inc.

NAME ______________________________ DATE ____________ PERIOD ______

6-2 Study Guide and Intervention

Ratio Tables

Lesson 6–2

A **ratio table** organizes data into columns that are filled with pairs of numbers that have the same ratio, or are equivalent. **Equivalent ratios** express the same relationship between two quantities.

Example 1 **BAKING You need 1 cup of rolled oats to make 24 oatmeal cookies. Use the ratio table at the right to find how many oatmeal cookies you can make with 5 cups of rolled oats.**

Cups of Oats	1				5
Oatmeal Cookies	24				■

Find a pattern and extend it.

		+1	+1	+1	+1
Cups of Oats	1	2	3	4	5
Oatmeal Cookies	24	48	72	96	120
		+24	+24	+24	+24

So, 120 oatmeal cookies can be made with 5 cups of rolled oats.

Multiplying or dividing two related quantities by the same number is called **scaling.** You may sometimes need to *scale back* and then *scale forward* or vice versa to find an equivalent ratio.

Example 2 **SHOPPING A department store has socks on sale for 4 pairs for $10. Use the ratio table at the right to find the cost of 6 pairs of socks.**

Pairs of Socks		4	6
Cost in Dollars		10	■

There is no whole number by which you can multiply 4 to get 6. Instead, scale back to 2 and then forward to 6.

So, the cost of 6 pairs of socks would be $15.

	÷ 2 (4 to 2)		× 3 (2 to 6)
Pairs of Socks	2	4	6
Cost in Dollars	5	10	15
	÷ 2 (10 to 5)		× 3 (5 to 15)

Exercises

1. **EXERCISE** Keewan bikes 6 miles in 30 minutes. At this rate, how long would it take him to bike 18 miles?

Distance Biked (mi)	6		18
Time (min)	30		■

2. **HOBBIES** Christine is making fleece blankets. 6 yards of fleece will make 2 blankets. How many blankets can she make with 9 yards of fleece?

Yards of Fleece		6	9
Number of Blankets		2	■

Copyright © Glencoe/McGraw-Hill, a division of The McGraw-Hill Companies, Inc.

6-2 Practice

Ratio Tables

For Exercises 1–3, use the ratio tables given to solve each problem.

1. **CAMPING** To disinfect 1 quart of stream water to make it drinkable, you need to add 2 tablets of iodine. How many tablets do you need to disinfect 4 quarts?

Number of Tablets	2		■
Number of Quarts	1		4

2. **BOOKS** A book store bought 160 copies of a book from the publisher for $4,000. If the store gives away 2 books, how much money will it lose?

Number of Copies	160		2
Cost in Dollars	4,000		■

3. **BIRDS** An ostrich can run at a rate of 50 miles in 60 minutes. At this rate, how long would it take an ostrich to run 18 miles?

Distance Run (mi)	50		18
Time (min)	60		■

4. **DISTANCE** If 10 miles is about 16 kilometers and the distance between two towns is 45 miles, use a ratio table to find the distance between the towns in kilometers. Explain your reasoning.

5. **SALARY** Luz earns $400 for 40 hours of work. Use a ratio table to determine how much she earns for 6 hours of work.

RECIPES For Exercises 6–8, use the following information.

A soup that serves 16 people calls for 2 cans of chopped clams, 4 cups of chicken broth, 6 cups of milk, and 4 cups of cubed potatoes.

6. Create a ratio table to represent this situation.

7. How much of each ingredient would you need to make an identical recipe that serves 8 people? 32 people?

8. How much of each ingredient would you need to make an identical recipe that serves 24 people? Explain your reasoning.

Copyright © Glencoe/McGraw-Hill, a division of The McGraw-Hill Companies, Inc.

NAME ______________________ DATE __________ PERIOD _____

6-3 Study Guide and Intervention

Proportions

Two quantities are said to be **proportional** if they have a constant ratio. A **proportion** is an equation stating that two ratios are equivalent.

Example 1 **Determine if the quantities in each pair of rates are proportional. Explain your reasoning and express each proportional relationship as a proportion.**

\$35 for 7 balls of yarn; \$24 for 4 balls of yarn.

Write each ratio as a fraction. Then find its unit rate.

$$\frac{\$35}{\text{7 balls of yarn}} \overset{\div 7}{\underset{\div 7}{=}} \frac{\$5}{\text{1 ball of yarn}} \qquad \frac{\$24}{\text{4 balls of yarn}} \overset{\div 4}{\underset{\div 4}{=}} \frac{\$6}{\text{1 ball of yarn}}$$

Since the ratios do not share the same unit rate, the cost is not proportional to the number of balls of yarn purchased.

Example 2 **Determine if the quantities in each pair of rates are proportional. Explain your reasoning and express each proportional relationship as a proportion.**

8 boys out of 24 students; 4 boys out of 12 students

Write each ratio as a fraction.

$$\frac{\text{8 boys}}{\text{24 students}} \overset{\div 2}{\underset{\div 2}{\quad}} \frac{\text{4 boys}}{\text{12 students}}$$ ←The numerator and the denominator are divided by the same number.

Since the fractions are equivalent, the number of boys is proportional to the number of students.

Exercises

Determine if the quantities in each pair of rates are proportional. Explain your reasoning and express each proportional relationship as a proportion.

1. \$12 saved after 2 weeks; \$36 saved after 6 weeks

2. \$9 for 3 magazines; \$20 for 5 magazines

3. 135 miles driven in 3 hours; 225 miles driven in 5 hours

4. 24 computers for 30 students; 48 computers for 70 students

Lesson 6–2

Copyright © Glencoe/McGraw-Hill, a division of The McGraw-Hill Companies, Inc.

NAME ______________________________ DATE ______________ PERIOD _____

6-3 Practice

Proportions

Determine if the quantities in each pair of ratios are proportional. Explain your reasoning and express each proportional relationship as a proportion.

1. 18 vocabulary words learned in 2 hours; 27 vocabulary words learned in 3 hours

2. $15 for 5 pairs of socks; $25 for 10 pairs of socks

3. 20 out of 45 students attended the concert; 12 out of 25 students attended the concert

4. 78 correct answers out of 100 test questions; 39 correct answers out of 50 test questions

5. 15 minutes to drive 21 miles; 25 minutes to drive 35 miles

ANIMALS For Exercises 6–8, refer to the table on lengths of some animals with long tails. Determine if each pair of animals has the same body length to tail length proportions. Explain your reasoning.

Animal Lengths (mm)		
Animal	**Head & Body**	**Tail**
Brown Rat	240	180
Hamster	250	50
Lemming	125	25
Opossum	480	360
Prairie Dog	280	40

6. brown rat and opossum

7. hamster and lemming

8. opossum and prairie dog

Copyright © Glencoe/McGraw-Hill, a division of The McGraw-Hill Companies, Inc.

NAME ______________________ DATE __________ PERIOD _____

6-4 Study Guide and Intervention

Algebra: Solving Proportions

To *solve a proportion* means to find the unknown value in the proportion. By examining how the numerators or denominators of the proportion are related, you can perform an operation on one fraction to create an equivalent fraction.

Example 1 **Solve $\frac{3}{4} = \frac{b}{12}$.**

Find a value for b that would make the fractions equivalent.

$$\frac{3}{4} \overset{\times 3}{=} \frac{b}{12}$$

Since 4 × 3 = 12, multiply the numerator and denominator by 3.

$b = 3 \times 3$ or 9

Example 2 **NUTRITION Three servings of broccoli contain 150 calories. How many servings of broccoli contain 250 calories?**

Set up the proportion. Let a represent the number of servings that contain 250 calories.

$$\frac{150 \text{ calories}}{3 \text{ servings}} = \frac{250 \text{ calories}}{a \text{ servings}}$$

Find the unit rate.

$$\frac{150 \text{ calories}}{3 \text{ servings}} \overset{\div 3}{=} \frac{50 \text{ calories}}{1 \text{ serving}}$$

Rewrite the proportion using the unit rate and solve using equivalent fractions.

$$\frac{50 \text{ calories}}{1 \text{ serving}} \overset{\times 5}{=} \frac{250 \text{ calories}}{5 \text{ servings}}$$

So, 5 servings of broccoli contain 250 calories.

Exercises

Solve each proportion.

1. $\frac{2}{3} = \frac{8}{n}$

2. $\frac{2}{4} = \frac{y}{8}$

3. $\frac{3}{5} = \frac{b}{15}$

4. $\frac{4}{5} = \frac{16}{w}$

5. $\frac{d}{16} = \frac{3}{8}$

6. $\frac{2}{y} = \frac{6}{9}$

7. **MUSIC** Jeremy spent $33 on 3 CDs. At this rate, how much would 5 CDs cost?

Lesson 6–4

Copyright © Glencoe/McGraw-Hill, a division of The McGraw-Hill Companies, Inc.

6-4 Practice

Algebra: Solving Proportions

Solve each proportion.

1. $\frac{2}{3} = \frac{n}{21}$

2. $\frac{2}{x} = \frac{16}{40}$

3. $\frac{80}{100} = \frac{b}{5}$

4. $\frac{m}{2} = \frac{75}{50}$

5. $\frac{6}{5} = \frac{42}{a}$

6. $\frac{3}{d} = \frac{21}{56}$

7. $\frac{4}{3} = \frac{f}{45}$

8. $\frac{h}{12} = \frac{70}{120}$

9. $\frac{3}{5} = \frac{27}{p}$

10. $\frac{26}{21} = \frac{r}{63}$

11. $\frac{17}{y} = \frac{102}{222}$

12. $\frac{7}{10} = \frac{c}{25}$

13. MAMMALS A pronghorn antelope can travel 105 miles in 3 hours. If it continued traveling at the same speed, how far could a pronghorn travel in 11 hours?

14. BIKES Out of 32 students in a class, 5 said they ride their bikes to school. Based on these results, predict how many of the 800 students in the school ride their bikes to school.

15. MEAT Hamburger sells for 3 pounds for $6. If Alicia buys 10 pounds of hamburger, how much will she pay?

16. FOOD If 24 extra large cans of soup will serve 96 people, how many cans should Ann buy to serve 28 people?

17. BIRDS The ruby throated hummingbird has a wing beat of about 200 beats per second. About how many wing beats would a hummingbird have in 3 minutes?

Copyright © Glencoe/McGraw-Hill, a division of The McGraw-Hill Companies, Inc.

NAME ______________________________ DATE ______________ PERIOD ______

6-5 Study Guide and Intervention

Problem-Solving Investigation: Look for a Pattern

When solving problems, one strategy that is helpful is to *look for a pattern.* In some problem situations, you can extend and examine a pattern in order to solve the problem.

You can use the *look for a pattern* strategy, along with the following four-step problem solving plan to solve a problem.

1 Understand – Read and get a general understanding of the problem.

2 Plan – Make a plan to solve the problem and estimate the solution.

3 Solve – Use your plan to solve the problem.

4 Check – Check the reasonableness of your solution.

Example MEDICINE **Monisha has the flu. The doctor gave her medicine to take over the next 2 weeks. The first 3 days she is to take 2 pills a day. Then the remaining days she is to take 1 pill. How many pills will Monisha have taken at the end of the 2 weeks?**

Understand You know she is to take the medicine for 2 weeks. You also know she is to take 2 pills the first 3 days and then only 1 pill the remaining days. You need to find the total number of pills.

Plan Start with the first week and look for a pattern.

Solve

Day	1	2	3	4	5	6	7
Number of Pills	2	2	2	1	1	1	1
Total Pills	2	$2 + 2 = 4$	$4 + 2 = 6$	$6 + 1 = 7$	$7 + 1 = 8$	$8 + 1 = 9$	$9 + 1 = 10$

After the first few days the number of pills increases by 1. You can add 7 more pills to the total for the first week, $10 + 7 = 17$. So, by the end of the 2 weeks, Monisha will have taken 17 pills to get over the flu.

Check You can extend the table for the next 7 days to check the answer.

Exercise

TIME Buses arrive every 30 minutes at the bus stop. The first bus arrives at 6:20 A.M. Hogan wants to get on the first bus after 8:00 A.M. What time will the bus that Hogan wants to take arrive at the bus stop?

Lesson 6–5

Copyright © Glencoe/McGraw-Hill, a division of The McGraw-Hill Companies, Inc.

NAME ______________________ DATE ____________ PERIOD ______

6-5 Practice

Problem-Solving Investigation: Look for a Pattern

Mixed Problem Solving

Use the look for a pattern strategy to solve Exercises 1 and 2.

1. **MONEY** In 2005, Trey had $7,200 in his saving-for-college account and Juan had $8,000. Each year, Trey will add $400 and Juan will add $200. In what year will they both have the same amount of money in their accounts, not counting interest earned? How much will it be?

2. **BUTTONS** Draw the next two figures in the pattern below.

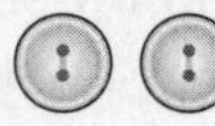

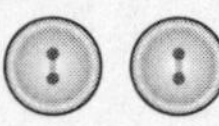

Use any strategy to solve Exercises 3–7. Some strategies are shown below.

Problem-Solving Strategies
• Guess and check.
• Look for a pattern.
• Act it out.

3. **MUSIC** Last week Jason practiced playing his bassoon for 95 minutes. This week he practiced 5 more minutes than 3 times the number of minutes he practiced last week. How many minutes did Jason practice this week?

4. **NUMBER SENSE** Describe the pattern below. Then find the missing number.

   5,000, 2,500, ■, 625, . . .

5. **TRAVEL** An express bus left the station at 6:30 a.m. and arrived at its destination at 12:00 noon. It traveled a distance of 260 miles and made only one stop for a half hour to drop off and pick up passengers. What was the average speed of the bus?

6. **MONEY** Len bought a $24.99 pair of pants and paid a total of $27.05, including tax. How much was the tax?

7. **PHOTOGRAPHY** Ms. Julian gives photography workshops. She collected $540 in fees for a workshop attended by 12 participants. Ms. Julian spent $15 per person for supplies for them and herself and $6 per person for box lunches for them and herself. How much money did Ms. Julian have left as profit?

Copyright © Glencoe/McGraw-Hill, a division of The McGraw-Hill Companies, Inc.

NAME ______________________ DATE ____________ PERIOD ______

6-6 Study Guide and Intervention

Sequences and Expressions

A **sequence** is a list of numbers in a specific order. Each number in the sequence is called a **term**. An **arithmetic sequence** is a sequence in which each term is found by adding the same number to the previous term.

Example **Use words and symbols to describe the value of each term as a function of its position. Then find the value of the tenth term in the sequence.**

Position	1	2	3	4	n
Value of Term	4	8	12	16	?

Study the relationship between each position and the value of its term.

Notice that the value of each term is 4 times its position number. So the value of the term in position n is $4n$.

To find the value of the tenth term, replace n with 10 in the algebraic expression $4n$. Since $4 \times 10 = 40$, the value of the tenth term in the sequence is 40.

Position		Value of term
1	$\times 4 =$	4
2	$\times 4 =$	8
3	$\times 4 =$	12
4	$\times 4 =$	16
n	$\times 4 =$	$4n$

Exercises

Use words and symbols to describe the value of each term as a function of its position. Then find the value of the tenth term in the sequence.

1.

Position	3	4	5	6	n
Value of Term	1	2	3	4	?

2.

Position	1	2	3	4	n
Value of Term	5	10	15	20	?

3.

Position	4	5	6	7	n
Value of Term	11	12	13	14	?

Copyright © Glencoe/McGraw-Hill, a division of The McGraw-Hill Companies, Inc.

Lesson 6–6

6-6 Practice

Sequences and Expressions

Use words and symbols to describe the value of each term as a function of its position. Then find the value of the sixteenth term in the sequence.

1.

Position	2	3	4	5	n
Value of Term	8	12	16	20	■

2.

Position	8	9	10	11	n
Value of Term	14	15	16	17	■

3.

Position	11	12	13	14	n
Value of Term	4	5	6	7	■

4.

Position	21	22	23	24	n
Value of Term	12	13	14	15	■

Determine how the next term in each sequence can be found. Then find the next two terms in the sequence.

5. 3, 16, 29, 42, …

6. 29, 25, 21, 17, …

7. 1.2, 3.5, 5.8, 8.1, …

Find the missing number in each sequence.

8. 5, ■, 10, $12\frac{1}{2}$, …

9. 11.5, 9.4, ■, 5.2

10. 40, ■, $37\frac{1}{3}$, 36, …

11. MEASUREMENT There are 52 weeks in 1 year. In the space at the right, make a table and write an algebraic expression relating the number of weeks to the number of years. Then find Hana's age in weeks if she is 11 years old.

12. COMPUTERS There are about 8 bits of digital information in 1 byte. In the space at the right, make a table and write an algebraic expression relating the number of bits to the number of bytes. Then find the number of bits there are in one kilobyte if there are 1,024 bytes in one kilobyte.

Copyright © Glencoe/McGraw-Hill, a division of The McGraw-Hill Companies, Inc.

NAME ______________________ DATE ____________ PERIOD ______

6-7 Study Guide and Intervention

Proportions and Equations

A *function table* displays *input* and *output* values that represent a function. The function displayed in a function table can be represented with an *equation*.

Example 1 **Write an equation to represent the function displayed in the table.**

Examine how the value of each input and output changes.

Input, x	1	2	3	4	5
Output, y	5	10	15	20	25

As each input increases by 1, the output increases by 5. That is, the constant rate of change is 5.

+1 +1 +1 +1

Input, x	1	2	3	4	5
Output, y	5	10	15	20	25

+5 +5 +5 +5

So, the equation that represents the function is $y = 5x$.

Example 2 **Theo earns \$6 an hour mowing lawns for his neighbors. Make a table and write an equation for the total amount *t* Theo earns for mowing *h* hours. How much will Theo earn for mowing lawns for 11 hours?**

As the number of hours increases by 1, the total earned increases by 6.

So, the equation is $t = 6h$.

Let $h = 11$ to find how much Theo will earn in 11 hours.

$t = 6h$
$t = 6 \times 11$ or \$66

Hours, h	Total earned, t
1	\$6
2	\$12
3	\$18
4	\$24

(+1 between each hours value; +6 between each total earned value)

Exercises

Write an equation to represent the function displayed in each table.

1.

Input, x	1	2	3	4	5
Output, y	2	4	6	8	10

2.

Input, x	0	1	2	3	4
Output, y	0	6	12	18	24

MUSIC Use the following information for Exercises 3–5.

A music store sells each used CD for \$4.

3. Make a table to show the relationship between the number of *c* used CDs purchased and the total cost *t*.

4. Write an equation to find *t*, the total cost in dollars for buying *c* used CDs.

5. How much will it cost to buy 5 used CDs?

Copyright © Glencoe/McGraw-Hill, a division of The McGraw-Hill Companies, Inc.

NAME ______________________________ DATE ____________ PERIOD _____

6-7 Practice

Proportions and Equations

Write an equation to represent the function displayed in each table.

1.

Input, x	1	2	3	4	5
Output, y	7	14	21	28	35

2.

Input, x	0	1	2	3	4
Output, y	0	9	18	27	36

3.

Input, x	1	2	3	4	5
Output, y	13	26	39	52	65

4.

Input, x	10	20	30	40	50
Output, y	1	2	3	4	5

5.

Input, x	0	1	2	3	4
Output, y	0	14	28	42	56

6.

Input, x	4	8	12	16	20
Output, y	1	2	3	4	5

7.

Input, x	12	24	36	48	60
Output, y	1	2	3	4	5

8.

Input, x	6	12	18	24	30
Output, y	1	2	3	4	5

BATS **Use the following information for Exercises 9–11.**

A Little Brown Myotis bat can eat 500 mosquitoes in an hour.

9. In the space at the right, make a table to show the the relationship between the number of hours h and the number of mosquitoes eaten m.

10. Write an equation to find m, the number of mosquitoes a bat eats in h hours.

11. How many mosquitoes can a Little Brown Myotis bat eat in 7 hours?

12. RECREATION A community center charges the amount shown in the table for using specialized exercise equipment. Write a sentence and an equation to describe the data. How much will it cost to use the exercise equipment for 6 months?

Number of Months, m	**Cost, c**
1	$20
2	$40
3	$60

Copyright © Glencoe/McGraw-Hill, a division of The McGraw-Hill Companies, Inc.

NAME ______________________ DATE ____________ PERIOD _____

7-1 Study Guide and Intervention

Percents and Fractions

To write a percent as a fraction, write it as a fraction with a denominator of 100. Then simplify.

Example 1 **Write 15% as a fraction in simplest form.**

15% means *15 out of 100*.

$15\% = \frac{15}{100}$ Definition of percent.

$= \frac{\overset{3}{\cancel{15}}}{\underset{20}{\cancel{100}}}$ or $\frac{3}{20}$ Simplify. Divide the numerator and denominator by the GCF, 5.

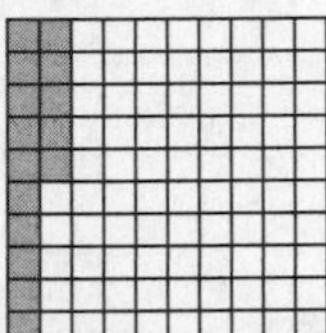

Example 2 **Write 180% as a fraction in simplest form.**

180% means *180 out of 100*.

$180\% = \frac{180}{100}$ Definition of percent.

$= \frac{\overset{9}{\cancel{180}}}{\underset{5}{\cancel{100}}}$ or $1\frac{4}{5}$ Simplify.

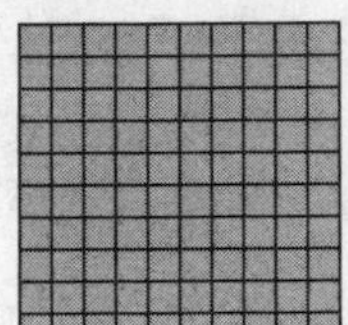

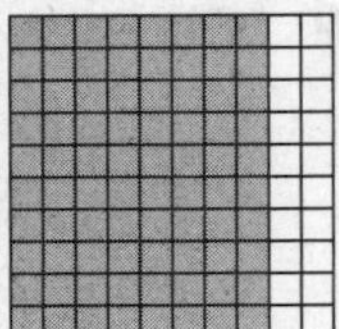

You can also write fractions as percents. To write a fraction as a percent, write a proportion and solve.

Example 3 **Write $\frac{2}{5}$ as a percent.**

$\frac{2}{5} = \frac{n}{100}$ Set up a proportion.

$\frac{2}{5} = \frac{40}{100}$ (× 20) Since 5 × 20 = 100, multiply 2 by 20 to find *n*.

So, $\frac{2}{5} = \frac{40}{100}$ or 40%

Example 4 **Write $\frac{7}{4}$ as a percent.**

$\frac{7}{4} = \frac{n}{100}$ Set up a proportion.

$\frac{7}{4} = \frac{175}{100}$ (× 25) Since 4 × 25 = 100, multiply 7 by 25 to find *n*.

So, $\frac{7}{4} = \frac{175}{100}$ or 175%.

Exercises

Write each percent as a fraction in simplest form.

1. 20% **2.** 35% **3.** 70%

4. 60% **5.** 150% **6.** 225%

Write each fraction as a percent.

7. $\frac{3}{10}$ **8.** $\frac{2}{100}$ **9.** $\frac{8}{5}$

10. $\frac{1}{5}$ **11.** $\frac{12}{5}$ **12.** $\frac{13}{100}$

Copyright © Glencoe/McGraw-Hill, a division of The McGraw-Hill Companies, Inc.

NAME ______________________ DATE ____________ PERIOD ______

7-1 Practice

Percents and Fractions

Write each percent as a fraction in simplest form.

1. 60% **2.** 18% **3.** 4%

4. 35% **5.** 10% **6.** 1%

7. 175% **8.** 258% **9.** 325%

10. **ENERGY** The United States uses 24% of the world's supply of energy. What fraction of the world's energy is this?

Write each fraction as a percent.

11. $\frac{6}{10}$ **12.** $\frac{2}{5}$ **13.** $\frac{9}{5}$

14. $\frac{6}{4}$ **15.** $\frac{7}{100}$ **16.** $\frac{4}{100}$

Write a percent to represent the shaded portion of each model.

17. 

18. 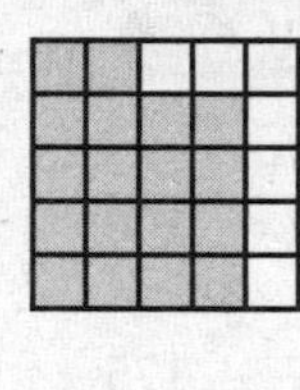

19. 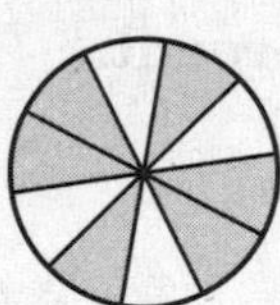

20. 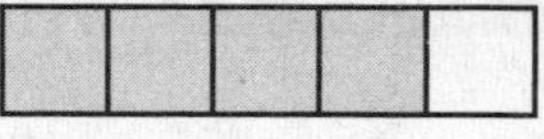

21. 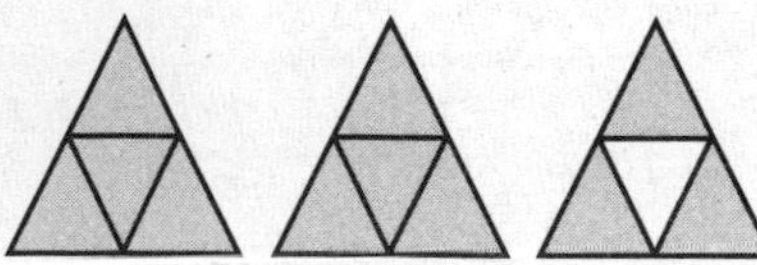

22. 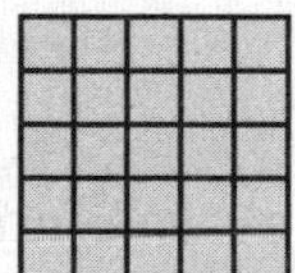 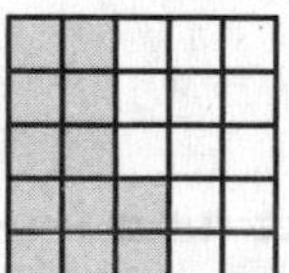

23. **ANALYZE TABLES** The table shows what fraction of a vegetable garden contains each kind of vegetable. What percent of the garden contains other kinds of vegetables?

Plant	Beans	Corn	Tomatoes	Other
Fraction	$\frac{1}{5}$	$\frac{1}{2}$	$\frac{1}{4}$	■

Copyright © Glencoe/McGraw-Hill, a division of The McGraw-Hill Companies, Inc.

NAME ______________________ DATE ____________ PERIOD ______

7-2 Study Guide and Intervention

Circle Graphs

Lesson 7–2

A **circle graph** is used to compare data that are parts of a whole. The pie-shaped sections show the groups. The percents add up to 100%.

Example 1 **The table shows the time Mike spends studying each subject during homework time. Sketch a circle graph of the data.**

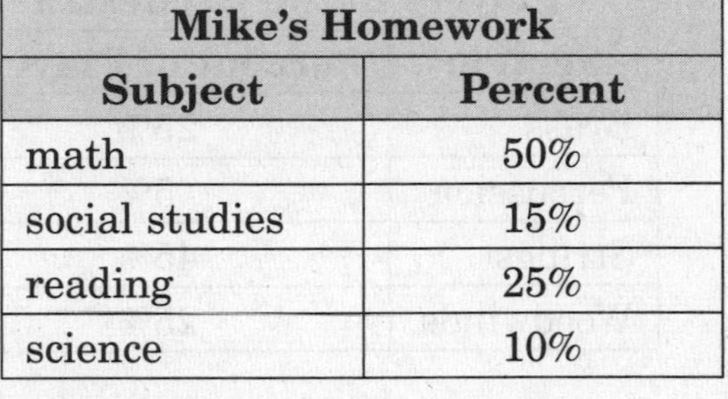

Mike's Homework	
Subject	**Percent**
math	50%
social studies	15%
reading	25%
science	10%

- Write each percent as a fraction.

$50\% = \frac{50}{100}$ or $\frac{1}{2}$ $15\% = \frac{15}{100}$ or $\frac{3}{20}$

$25\% = \frac{25}{100}$ or $\frac{1}{4}$ $10\% = \frac{10}{100}$ or $\frac{1}{10}$

- Use a compass to draw a circle.
- Since $50\% = \frac{1}{2}$, shade and label $\frac{1}{2}$ of the circle for math.

Since $25\% = \frac{1}{4}$, shade and label $\frac{1}{4}$ of the circle for reading.

Split the remaining section so that one section is slightly larger than the other. Label the slightly larger section social studies for 15%, and the smaller one science for 10%.

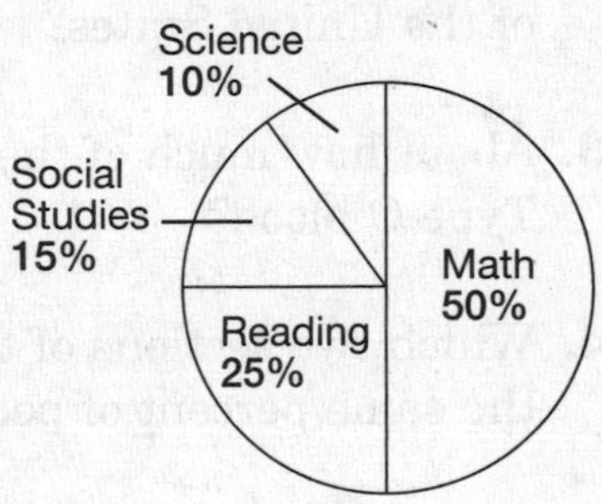

Example 2 **In the circle graph to the right, how does the amount of time Mike spends studying math compare to the amount of time he studies reading?**

The section representing math is twice the size of the section representing reading. So, Mike spends twice as much time studying math as reading.

Exercises

SURVEYS Use the table that shows the results of a favorite colors survey.

Favorite Color	
Color	**Percent**
blue	33%
red	25%
green	25%
purple	10%
yellow	7%

1. Sketch a circle graph of the data.

2. In your circle graph, which two sections represent the responses by the same amount of students?

3. In your circle graph, how does the number of students that chose blue compare to the number of students that chose purple?

Copyright © Glencoe/McGraw-Hill, a division of The McGraw-Hill Companies, Inc.

7-2 Practice

Circle Graphs

1. **MUSIC** The table shows the percent of students in the school orchestra who played in each section. Sketch a circle graph to display the data.

Players in the Orchestra	
Section	**Percent of Players**
Brass	25%
Percussion	5%
Strings	45%
Woodwinds	25%

BLOOD For Exercises 2–5, use the graph that shows the percent of Americans having different blood types.

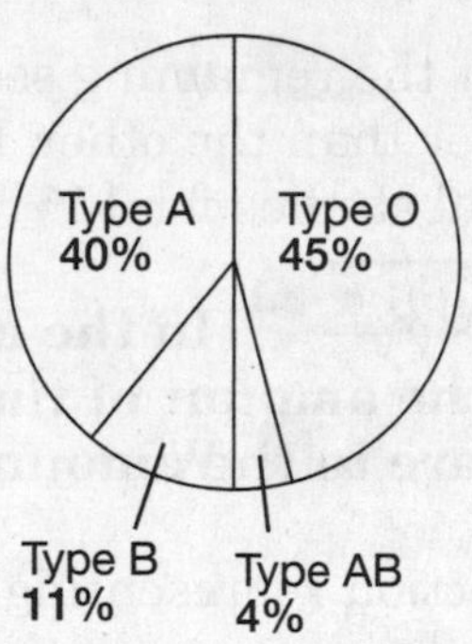

2. Which blood type is the least common among people of the United States?

3. About how much of the total U.S. population has Type O blood?

4. Which two sections of the graph represent about the same percent of people? Explain your reasoning.

5. How does Type A compare to Type AB in number of people having these two types?

6. **FOOD** A group of 100 students were asked about their favorite sandwiches. The chart shows their responses. In the space at the right, sketch a circle graph to compare the students' responses. What percent of the students chose luncheon meat or tuna as their favorite sandwich?

Favorite Sandwich	
Response	**Number of Students**
Egg Salad	12
Luncheon Meat	51
Nut Butter and Jelly	24
Tuna	11
Other	2

Copyright © Glencoe/McGraw-Hill, a division of The McGraw-Hill Companies, Inc.

NAME ______________________ DATE __________ PERIOD _____

7-3

Study Guide and Intervention

Percents and Decimals

To write a percent as a decimal, first rewrite the percent as a fraction with a denominator of 100. Then write the fraction as a decimal.

Example 1 **Write 23% as a decimal.**

$23\% = \frac{23}{100}$ Rewrite the percent as a fraction with a denominator of 100.

$= 0.23$ Write the fraction as a decimal.

Example 2 **Write 127% as a decimal.**

$127\% = \frac{127}{100}$ Rewrite the percent as a fraction with a denominator of 100.

$= 1.27$ Write the fraction as a decimal.

To write a decimal as a percent, first write the decimal as a fraction with a denominator of 100. Then write the fraction as a percent.

Example 3 **Write 0.44 as a percent.**

$0.44 = \frac{44}{100}$ Write the decimal as a fraction.

$= 44\%$ Write the fraction as a percent.

Example 4 **Write 2.65 as a percent.**

$2.65 = 2\frac{65}{100}$ Write *2 and 65 hundredths* as a mixed number.

$= \frac{265}{100}$ Write the mixed number as an improper fraction.

$= 265\%$ Write the fraction as a percent.

Exercises

Write each percent as a decimal.

1. 39% **2.** 57% **3.** 82%

4. 135% **5.** 112% **6.** 0.4%

Write each decimal as a percent.

7. 0.86 **8.** 0.36 **9.** 0.65

10. 0.2 **11.** 1.48 **12.** 2.17

Lesson 7–3

Copyright © Glencoe/McGraw-Hill, a division of The McGraw-Hill Companies, Inc.

NAME ______________________________ DATE ____________ PERIOD _____

7-3 Practice

Percents and Decimals

Express each percent as a decimal.

1. 29% **2.** 63% **3.** 4% **4.** 9%

5. 148% **6.** 106% **7.** 10% **8.** 32%

9. **ENERGY** The United States gets about 39% of its energy from petroleum. Write 39% as a decimal.

10. **SCIENCE** About 8% of the earth's crust is made up of aluminum. Write 8% as a decimal.

Express each decimal as a percent.

11. 0.45 **12.** 0.12 **13.** 1.68 **14.** 2.73

15. 0.2 **16.** 0.7 **17.** 0.95 **18.** 0.46

19. **POPULATION** In 2000, the number of people 65 years and older in Arizona was 0.13 of the total population. Write 0.13 as a percent.

20. **GEOGRAPHY** About 0.41 of Hawaii's total area is water. What percent is equivalent to 0.41?

Replace each ● with <, >, or = to make a true sentence.

21. 26% ● 0.3 **22.** 0.9 ● 9% **23.** 4.7 ● 47%

24. **ANALYZE TABLES** A batting average is the ratio of hits to at bats. Batting averages are expressed as a decimal rounded to the nearest thousandth. Show two different ways of finding how much greater Derek Jeter's batting average was than Jason Giambi's batting average. Express as a percent.

New York Yankees, 2005 Batting Statistics	
Player	**Batting Average**
Jason Giambi	0.286
Derek Jeter	0.307
Hideki Matsui	0.297
Jorge Posada	0.257

Source: ESPN

Copyright © Glencoe/McGraw-Hill, a division of The McGraw-Hill Companies, Inc.

NAME ______________________ DATE __________ PERIOD ______

7-4 Study Guide and Intervention

Probability

When tossing a coin, there are two possible **outcomes**, heads and tails. Suppose you are looking for heads. If the coin lands on heads, this would be a favorable outcome or **simple event**. The chance that some event will happen (in this case, getting heads) is called **probability**. You can use a ratio to find probability. The probability of an event is a number from 0 to 1, including 0 and 1. The closer a probability is to 1, the more likely it is to happen.

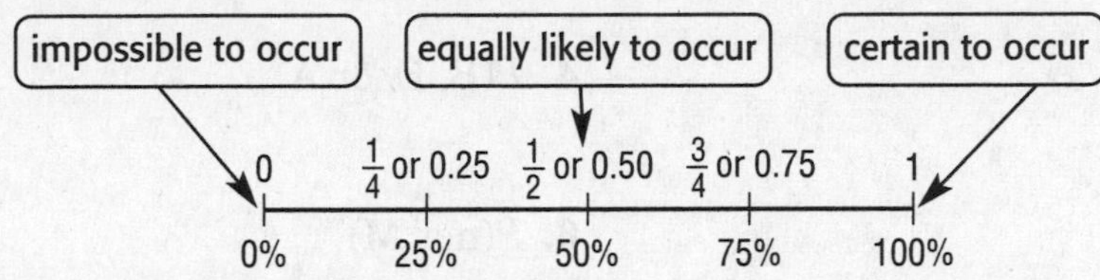

Example 1 **There are four equally likely outcomes on the spinner. Find the probability of spinning green or blue.**

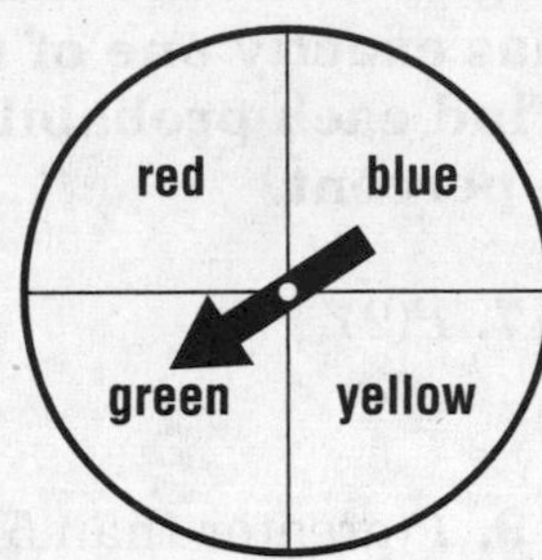

$$P(\text{green or blue}) = \frac{\text{number of favorable outcomes}}{\text{number of possible outcomes}}$$

$$= \frac{2}{4} \text{ or } \frac{1}{2}$$

The probability of landing on green or blue is $\frac{1}{2}$, 0.50, or 50%.

Complementary events are two events in which either one or the other must happen, but both cannot happen at the same time. The sum of the probabilities of complementary events is 1.

Example 2 **There is a 25% chance that Sam will win a prize. What is the probability that Sam will not win a prize?**

$P(\text{win}) + P(\text{not win}) = 1$

$0.25 + P(\text{not win}) = 1$ Replace P(win) with 0.25.

$-0.25 \quad = -0.25$ Subtract 0.25 from each side.

$P(\text{not win}) = 0.75$

So, the probability that Sam won't win a prize is 0.75, 75%, or $\frac{3}{4}$.

Exercises

1. There is a 90% chance that it will rain. What is the probability that it will not rain?

One pen is chosen without looking from a bag that has 3 blue pens, 6 red, and 3 green. Find the probability of each event. Write each answer as a fraction, a decimal, and a percent.

2. P(green)

3. P(blue or red)

4. P(yellow)

Lesson 7–4

Copyright © Glencoe/McGraw-Hill, a division of The McGraw-Hill Companies, Inc.

NAME ______________________ DATE ____________ PERIOD _____

7-4 Practice
Probability

The spinner shown is spun once. Find each probability. Write each answer as a fraction, a decimal, and a percent.

1. *P*(C)

2. *P*(G)

3. *P*(M or P)

4. *P*(B, E, or A)

5. *P*(not vowel)

6. *P*(not M)

Eight cards are marked 3, 4, 5, 6, 7, 8, 9, and 10 such that each card has exactly one of these numbers. A card is picked without looking. Find each probability. Write each answer as a fraction, a decimal, and a percent.

7. *P*(9)

8. *P*(5 or 7)

9. *P*(greater than 5)

10. *P*(less than 3)

11. *P*(odd)

12. *P*(4, 7, or 8)

13. *P*(not 6)

14. *P*(not 5 and not 10)

The spinner is spun once. Write a sentence stating how likely it is for each event to happen. Justify your answer.

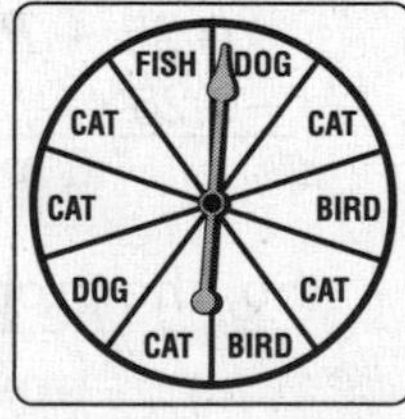

15. fish

16. cat

17. bird, cat, or fish

18. **PLANTS** Of the water lilies in the pond, 43% are yellow. The others are white. A frog randomly jumps onto a lily. Describe the complement of the frog landing on a yellow lily and find its probability.

Copyright © Glencoe/McGraw-Hill, a division of The McGraw-Hill Companies, Inc.

NAME ______________________________ DATE ____________ PERIOD _____

7-5 Study Guide and Intervention

Constructing Sample Spaces

The **Fundamental Counting Principle** is another way to find the number of possible outcomes. This principle states that if there are *m* outcomes for a first choice and *n* outcomes for a second choice, then the total number of possible outcomes can be found by finding $m \times n$.

Example 1 **How many sandwiches are possible from a choice of turkey or ham with jack cheese or Swiss cheese?**

Draw a tree diagram.

Sandwich	Cheese	Outcome
turkey (T)	jack (J)	TJ
	Swiss (S)	TS
ham (H)	jack (J)	HJ
	Swiss (S)	HS

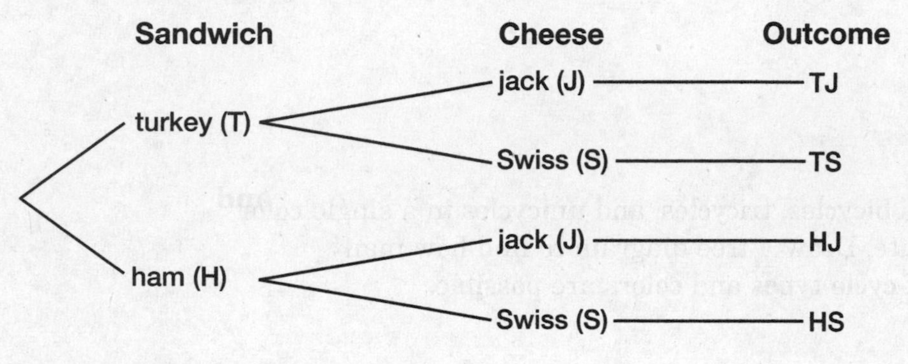

There are four possible sandwiches.

Example 2 **Using the Fundamental Counting Principle, how many sandwiches are possible from a choice of roast beef, turkey, or ham, with a choice of jack, cheddar, American, or Swiss cheese? Find the probability of chossing a ham with jack cheese sandwich.**

There are twelve possible sandwiches. To determine the number of possible outcomes, multiply the number of first choices, 3, by the number of second choices, 4, to determine that there are 12 possible outcomes. So, $P(\text{ham, jack}) = \frac{1}{12}$, or 0.083, or 8.3%.

Exercises

First use the Fundamental Counting Principle to determine the number of possible outcomes. Then, check your result and find the sample space by drawing a tree diagram. Finally, find the probability.

1. buy a can or a bottle of grape or orange soda
   Find *P*(bottle, grape).

2. toss a coin and roll a number cube
   Find *P*(4, tails).

3. wear jeans or shorts with a blue, white, black, or red T-shirt. Find *P*(jeans, white T-shirt).

Lesson 7–5

Copyright © Glencoe/McGraw-Hill, a division of The McGraw-Hill Companies, Inc.

NAME ________________________ DATE ____________ PERIOD _____

7-5 Practice

Constructing Sample Spaces

1. **SCULPTURE** Diego is lining up driftwood sculptures in front of his woodshop. He has a dolphin, gull, seal, and a whale. In how many different ways can he line up his sculptures? Make an organized list to show the sample space.

2. **CYCLES** A cycle shop sells bicycles, tricycles, and unicycles in a single color of red, blue, green, or white. Draw a tree diagram to find how many different combinations of cycle types and colors are possible.

For Exercises 3–5, a coin is tossed, and the spinners shown are spun.

Spinner 1

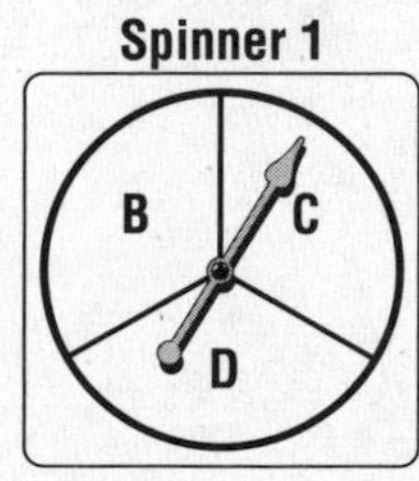

Spinner 2

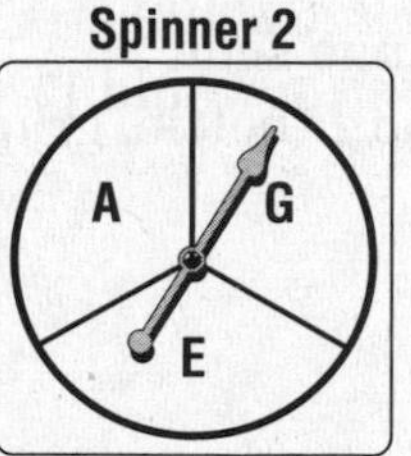

3. Using the Fundamental Counting Principle, how many outcomes are possible?

4. What is *P*(heads, C, G)?

5. Find *P*(tails, D, a vowel).

Copyright © Glencoe/McGraw-Hill, a division of The McGraw-Hill Companies, Inc.

NAME ______________________ DATE __________ PERIOD _____

7-6 Study Guide and Intervention

Making Predictions

A **survey** is a method of collecting information. The group being surveyed is the **population**. To save time and money, part of the group, called a **sample**, is surveyed.

A good sample is:

- selected at **random**, or without preference,
- representative of the population, and
- large enough to provide accurate data.

Examples **Every sixth student who walked into the school was asked how he or she got to school.**

1 What is the probability that a student at the school rode a bike to school?

School Transportation	
Method	**Students**
walk	10
ride bike	10
ride bus	15
get ride	5

$$P(\text{ride bike}) = \frac{\text{number of students that rode a bike}}{\text{number of students surveyed}}$$

$$= \frac{10}{40} \text{ or } \frac{1}{4}$$

So, $P(\text{ride bike}) = \frac{1}{4}$, 0.25, or 25%.

2 There are 360 students at the school. Predict how many bike to school.

Write a proportion. Let s = number of students who will ride a bike.

$$\frac{10}{40} = \frac{s}{360}$$

You can solve the proportion to find that of the 360 students, 90 will ride a bike to school.

Exercises

SCHOOL **Use the following information and the table shown. Every tenth student entering the school was asked which one of the four subjects was his or her favorite.**

Favorite Subject	
Subject	**Students**
Language Arts	10
Math	10
Science	15
Social Studies	5

1. Find the probability that any student attending school prefers science.

2. There are 400 students at the school. Predict how many students would prefer science.

Copyright © Glencoe/McGraw-Hill, a division of The McGraw-Hill Companies, Inc.

NAME ______________________________ DATE ____________ PERIOD _____

7-6 Practice

Making Predictions

QUIZ SHOW For Exercises 1 and 2, use the following information.

On a quiz show, a contestant correctly answered 9 of the last 12 questions.

1. Find the probability of the contestant correctly answering the next question.

2. Suppose the contestant continues on the show and tries to correctly answer 24 questions. About how many questions would you predict the contestant to correctly answer?

CHORES For Exercises 3–6, use the table to predict the number of students out of 528 that would say each of the following was their least favorite chore.

Least Favorite Chore	
Chore	**Number of Students**
Clean my room	7
Take out the garbage	4
Wash dishes	5
Walk the dog	3
Vacuum or dust	5

3. clean my room

4. wash dishes

5. walk the dog

6. take out the garbage

7. **SCIENCE** Refer to the bar graph below. A science museum manager asked some of the visitors at random during a typical day which exhibit they preferred. If there are 630 visitors on a typical day, predict the number of visitors who prefer the magnets exhibit. Compare this to the number of visitors who prefer the weather exhibit.

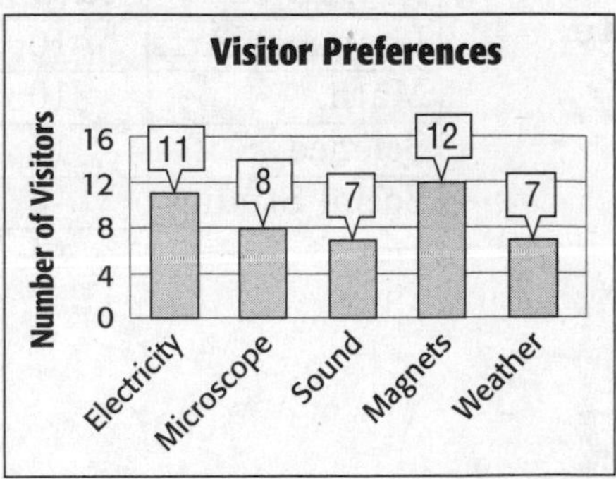

Copyright © Glencoe/McGraw-Hill, a division of The McGraw-Hill Companies, Inc.

7-7 Study Guide and Intervention

Problem-Solving Investigation: Solve a Simpler Problem

When solving problems, one strategy that is helpful is to *solve a simpler problem*. Using some of the information presented in the problem, you may be able to set up and solve a simpler problem.

You can use the *solve a simpler problem* strategy, along with the following four-step problem solving plan to solve a problem.

1 Understand – Read and get a general understanding of the problem.

2 Plan – Make a plan to solve the problem and estimate the solution.

3 Solve – Use your plan to solve the problem.

4 Check – Check the reasonableness of your solution.

Example PUZZLES **Steven and Darshelle are putting together a 500 piece puzzle. So far they have 40% of the puzzle complete. How many pieces are left for them to fit into the puzzle?**

Understand We know the total number of pieces in the puzzle and that 40% of the pieces are already put together in the puzzle. We need to find the number of pieces left to fit in the puzzle.

Plan Solve a simpler problem by finding 100% − 40% or 60% of the 500 pieces. First find 10% of 500 and then use the result to find 60% of 500.

Solve Since 10%, or $\frac{1}{10}$ of 500 is 50.

So, 60%, or $\frac{6}{10}$ of 500 is 6×50 or 300.

Steven and Darshelle still have 300 pieces left to fit in the puzzle.

Check We know that 40% or 4 out of every 10 pieces of the puzzle are already put together in the puzzle. Since $500 \div 10 \times 4 = 200$ pieces and $200 + 300 = 500$, the answer is correct.

Exercise

SCHOLARSHIPS Crosswood Elementary School received $450 in donations for its scholarship fund. If 30% of the contributions were from local businesses, how much money did the local businesses contribute?

Copyright © Glencoe/McGraw-Hill, a division of The McGraw-Hill Companies, Inc.

NAME ______________________ DATE ____________ PERIOD _____

7-7 Practice

Problem-Solving Investigation: Solve a Simpler Problem

Mixed Problem Solving

Use the solve a simpler problem strategy to solve Exercises 1–3.

1. **ART** An artist plans to make 1 clay pot the first week and triple the number of clay pots each week for 5 weeks. How many clay pots will the artist make the fifth week?

2. **GEOGRAPHY** The total area of Wisconsin is 65,498 square miles. Of that, about 80% is land area. About how much of Wisconsin is not land area?

3. **SCIENCE** Sound travels through sea water at a speed of about 1,500 meters per second. At this rate, how far will sound travel in 2 minutes?

Use any strategy to solve Exercises 4–8. Some strategies are shown below.

Problem-Solving Strategies
• Guess and check.
• Solve a simpler problem.

4. **MUSIC** Tanya scored 50 out of 50 points in her latest piano playing evaluation. She scored 42, 48, and 45 on previous evaluations. What score does she need on the next evaluation to have an average score of 45?

5. **EXERCISE** At the community center, 9 boys and 9 girls are playing singles table tennis. If each girl plays against each boy exactly once, how many games are played?

6. **CLOCK** The clock in the bell tower rings every half hour. How many times will it ring in one week?

7. **VENN DIAGRAMS** The Venn diagram shows information about the sixth graders in the school.

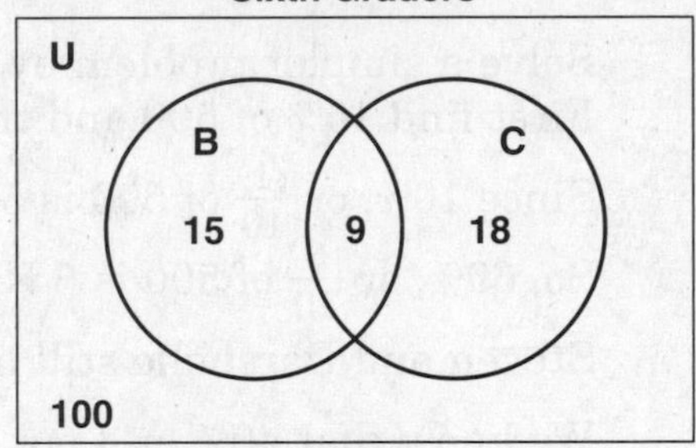

U = all sixth graders
B = sixth graders in the band
C = sixth graders in the chorus

How many more sixth graders in the school do not participate in band or chorus than do participate in band or chorus?

8. **MONEY** Kono wants to give $69 to charity. He will give each of 3 charities an equal amount of money. How much money will each charity receive?

Copyright © Glencoe/McGraw-Hill, a division of The McGraw-Hill Companies, Inc.

NAME ______________________ DATE ____________ PERIOD _____

7-8 Study Guide and Intervention

Estimating with Percents

The table below shows some commonly used percents and their fraction equivalents.

Percent-Fraction Equivalents				
$20\% = \frac{1}{5}$	$50\% = \frac{1}{2}$	$80\% = \frac{4}{5}$	$25\% = \frac{1}{4}$	$33\frac{1}{3}\% = \frac{1}{3}$
$30\% = \frac{3}{10}$	$60\% = \frac{3}{5}$	$90\% = \frac{9}{10}$	$75\% = \frac{3}{4}$	$66\frac{2}{3}\% = \frac{2}{3}$
$40\% = \frac{2}{5}$	$70\% = \frac{7}{10}$	$100\% = 1$		

Examples **Estimate each percent.**

 **20% of 58**

20% is $\frac{1}{5}$.

Round 58 to 60 since it is divisible by 5.

$\frac{1}{5}$ of 60 is 12.

So, 20% of 58 is about 12.

2 76% of 21.

76% is close to 75% or $\frac{3}{4}$.

Round 21 to 20 since it is divisible by 4.

$\frac{1}{4}$ of 24 is 5.

So $\frac{3}{4}$ of 20 is 3×5 or 15

So, 76% of 21 is about 15.

Example 3 **Isabel is reading a book that has 218 pages. She wants to complete 25% of the book by Friday. About how many pages should she read by Friday?**

25% os $\frac{1}{4}$. Round 218 to 200.

$\frac{1}{4}$ of 200 is 50.

So, Isabel should read about 50 pages by Friday.

Exercises

Estimate each percent.

1. 49% of 8

2. 24% of 27

3. 19% of 46

4. 62% of 20

5. 40% of 51

6. 81% of 32

7. **TIPS** Jodha wants to tip the pizza delivery person about 20%. If the cost of the pizzas is $15.99, what would be a reasonable amount to tip?

Lesson 7–8

Copyright © Glencoe/McGraw-Hill, a division of The McGraw-Hill Companies, Inc.

NAME ______________________ DATE ____________ PERIOD _____

7-8 Practice

Estimating with Percents

Estimate each percent.

1. 51% of 62 **2.** 39% of 42 **3.** 78% of 148 **4.** 34% of 99

5. 74% of 238 **6.** 70% of 103 **7.** 22% of 152 **8.** 91% of 102

9. 26% of 322 **10.** 65% of 181 **11.** 98% of 60 **12.** 11% of 10

13. Estimate twenty-nine percent of forty-eight.

14. Estimate sixty-two percent of one hundred twenty-four.

Estimate the percent that is shaded in each figure.

15. 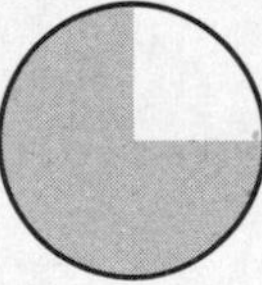

16. 

17. 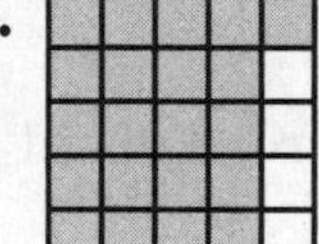

18. **WORK** Karl made $365 last month doing odd jobs after school. If 75% of the money he made was from doing yard work, about how much did Karl make doing yard work?

19. **HOMEWORK** Jin spent 32 hours on math and language arts homework last month. She spent 11 hours on math. About what percent of her homework hours were spent on language arts? Explain.

Copyright © Glencoe/McGraw-Hill, a division of The McGraw-Hill Companies, Inc.

8-1 Study Guide and Intervention

Length in the Customary System

The most commonly used customary units of length are shown in the table.

Customary Units Of Length	
Unit	**Model**
1 **inch** (in.)	width of a quarter
1 **foot** (ft) = 12 in.	length of a large adult foot
1 **yard** (yd) = 3 ft	length from nose to fingertip
1 **mile** (mi) = 1,760 yd	10 city blocks

Most rulers are divided into eighths of an inch, so you can measure to the nearest eighth inch.

Example 1 **Draw a line segment measuring $1\frac{5}{8}$ inches.**

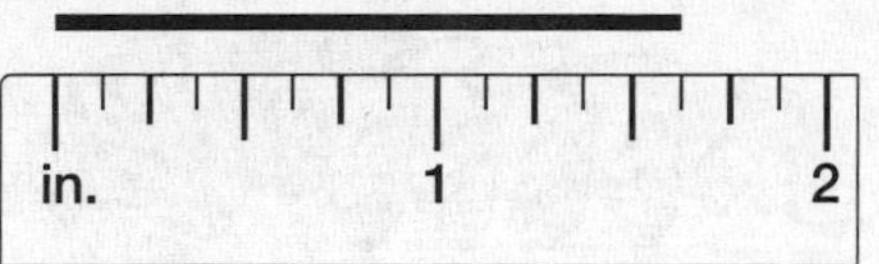

Draw a line segment from 0 to $1\frac{5}{8}$.

Example 2 **Measure the length of the nail to the nearest half, fourth, or eighth inch.**

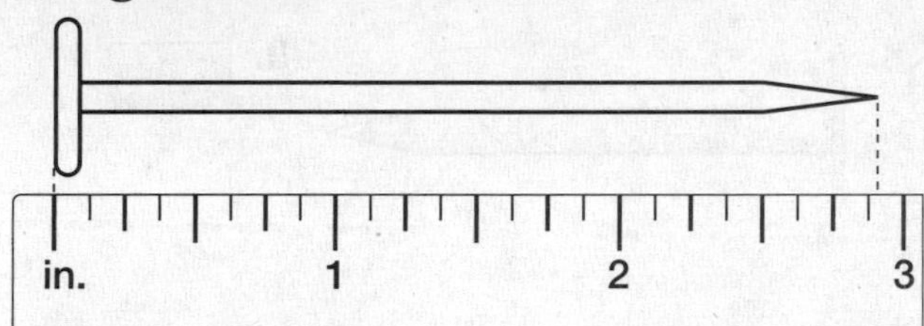

The nail is between $2\frac{7}{8}$ inches and 3 inches.
It is closer to $2\frac{7}{8}$ inches.
The length of the nail is about $2\frac{7}{8}$ inches.

- To change from larger units of length to smaller units, multiply.
- To change from smaller units of length to larger units, divide.

Examples **Complete.**

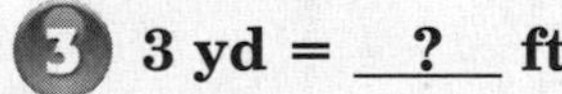

3 **3 yd = __?__ ft**

Since 1 yard = 3 feet, multiply by 3.
$3 \times 3 = 9$
So, 3 yards = 9 feet.

4 **24 in. = __?__ ft**

Since 1 foot = 12 inches, divide by 12.
$24 \div 12 = 2$
So, 24 inches = 2 feet.

Exercises

1. Draw a line segment that is $\frac{3}{4}$ in. long.

2. Measure the length of the object to the nearest half, fourth, or eighth inch.

Complete.

3. 3 ft = __?__ in. **4.** 15 ft = __?__ yd **5.** 2 mi = __?__ yd

Copyright © Glencoe/McGraw-Hill, a division of The McGraw-Hill Companies, Inc.

8-1 Practice

Length in the Customary System

Draw a line segment of each length.

1. $\frac{5}{8}$ in.

2. $3\frac{3}{8}$ in.

3. $2\frac{1}{4}$ in.

4. $1\frac{1}{2}$ in.

Measure the length of each line segment or object to the nearest half, fourth, or eighth inch.

5. •————•   **6.** •————•   **7.** •————•

8. 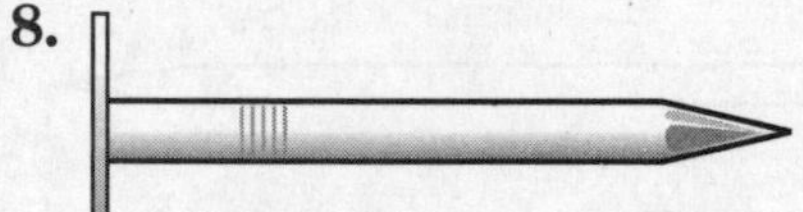

9. 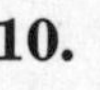

10. 

Complete.

11. 5 yd = _____ ft   **12.** 3 yd = _____ in.   **13.** 6 mi = _____ yd

14. 4 mi = _____ ft   **15.** 72 in. = _____ ft   **16.** 8 ft = _____ yd

17. 48 in. = _____ yd   **18.** 9,680 yd = _____ mi   **19.** 15,840 ft = _____ mi

Determine the greater measurement. Explain your reasoning.

20. $1\frac{2}{3}$ yards or 64 inches

21. 58 inches or $5\frac{1}{2}$ feet

22. **HUMAN BODY** The small intestine is 20 feet long and the large intestine is 5 feet long. What is the total length of the intestines in yards?

23. **BICYCLES** Raj estimates that the length of his bicycle is 66 inches. Is this reasonable? Why or why not?

Copyright © Glencoe/McGraw-Hill, a division of The McGraw-Hill Companies, Inc.

NAME ______________________ DATE ____________ PERIOD _____

8-2 Study Guide and Intervention

Capacity and Weight in the Customary System

The most commonly used customary units of capacity are shown below.

Customary Units Of Capacity	
Unit	**Model**
1 **fluid ounce** (fl oz)	2 tablespoons of water
1 **cup** (c) = 8 fl oz	coffee cup
1 **pint** (pt) = 2 c	small ice cream container
1 **quart** (qt) = 2 pt	large measuring cup
1 **gallon** (gal) = 4 qt	large plastic jug of milk

- To change from larger units of length to smaller units, multiply.
- To change from smaller units of length to larger units, divide.

Example 1 **Complete.**

2 gal = __?__ qt — THINK 1 gallon = 4 quarts

$2 \times 4 = 8$ — Multiply to change a larger unit to a smaller unit.

So, 2 gallons = 8 quarts.

The most commonly used customary units of weight are shown below.

Customary Units Of Weight	
Unit	**Model**
1 **ounce** (oz)	pencil
1 **pound** (lb) = 16 oz	package of notebook paper
1 **ton** (T) = 2,000 lb	small passenger car

Example 2 FOOD **A box of cereal weighs 32 ounces. How many pounds is this?**

32 oz = __?__ lb — THINK 16 ounces = 1 pound

$32 \div 16 = 2$ — Divide to change ounces to pounds.

So, 32 ounces = 2 pounds.

Exercises

Complete.

1. 2 pt = __?__ c **2.** 32 fl oz = __?__ c **3.** 3 lb = __?__ oz

4. 16 qt = __?__ gal **5.** 3 qt = __?__ pt **6.** 3 T = __?__ lb

7. 16 c = __?__ qt **8.** 2 gal = __?__ pt **9.** 64 oz = __?__ lb

Lesson 8–2

Copyright © Glencoe/McGraw-Hill, a division of The McGraw-Hill Companies, Inc.

NAME ______________________ DATE ____________ PERIOD _____

8-2 Practice

Capacity and Weight in the Customary System

Complete.

1. 6 gal = _____ qt **2.** 4 pt = _____ c **3.** 32 fl oz = _____ c

4. 9 qt = _____ pt **5.** 15 qt = _____ gal **6.** 7 gal = _____ pt

7. 3,000 lb = _____ T **8.** 68 oz = _____ lb **9.** 7 T = _____ lb

Write and solve a proportion to complete each conversion.

10. 9 qt = _____ pt **11.** 5 lb = _____ oz **12.** 56 fl oz = _____ c

Choose the better estimate for each measure.

13. fluid ounces or pints

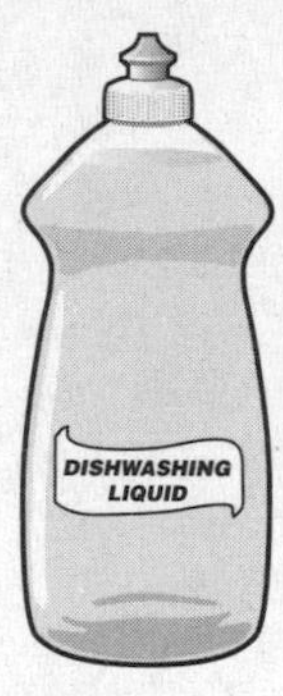

14. ounces or pounds

15. pints or gallons

Find the greater quantity. Explain your reasoning.

16. 18 quarts or 4 gallons

17. 3 pints or 36 fluid ounces

18. FISH The average weight of the largest fish, the whale shark, is 50,000 pounds. How many tons is this?

19. HONEY To gather enough nectar to make 1 pound of honey, a bee must visit 2 million flowers. How many flowers must a bee visit to make 64 ounces of honey?

20. PAINTING Mr. Krauss needs 8 gallons of paint to paint his fences. He has 9 quart cans and 22 pint cans of paint. Does he have enough paint? Explain.

Copyright © Glencoe/McGraw-Hill, a division of The McGraw-Hill Companies, Inc.

NAME ______________________ DATE ____________ PERIOD ____

8-3 Study Guide and Intervention

Length in the Metric System

The meter is the basic unit of length in the metric system. The most commonly used metric units of length are shown below.

Metric Units of Length		
Unit	**Model**	**Benchmark**
1 **millimeter** (mm)	thickness of a dime	25 mm ≈ 1 inch
1 **centimeter** (cm)	half the width of a penny	2.5 cm ≈ 1 inch
1 **meter** (m)	width of a doorway	1 m ≈ 1.1 yard
1 **kilometer** (km)	six city blocks	1.6 km ≈ 1 mile

Examples **Write the metric unit of length that you would use to measure each of the following.**

1 height of a box of popcorn

The height of a box of popcorn is more than the width of a penny, but less than the width of a doorway. So, the centimeter is an appropriate unit of measure.

2 length of a car

Since the length of a car is greater than the width of a doorway, but less than six city blocks, the meter is an appropriate unit of measure.

Example 3 **Measure the length of the line segment in centimeters.**

5 cm

cm 1 2 3 4 5

The line segment is 5 cm.

Exercises

Write the metric unit of length that you would use to measure each of the following.

1. height of a mountain

2. thickness of a dried bean

3. length of a pen

4. height of a table

Measure each line segment in centimeters and millimeters.

5. ________

6. ____________

7. ________________

8. ___

Copyright © Glencoe/McGraw-Hill, a division of The McGraw-Hill Companies, Inc.

NAME ______________________________ DATE ____________ PERIOD _____

8-3 Practice

Length in the Metric System

Write the metric unit of length that you would use to measure each of the following.

1. length of a fly

2. thickness of a pen

3. length of a football field

4. width of a sheet of notebook paper

5. width of a drinking glass

6. height of a mountain

7. distance from New York City to Los Angeles, California

8. distance from one end of a school to the other end of the school

Estimate the length of each of the following. Then measure to find the actual length.

9.

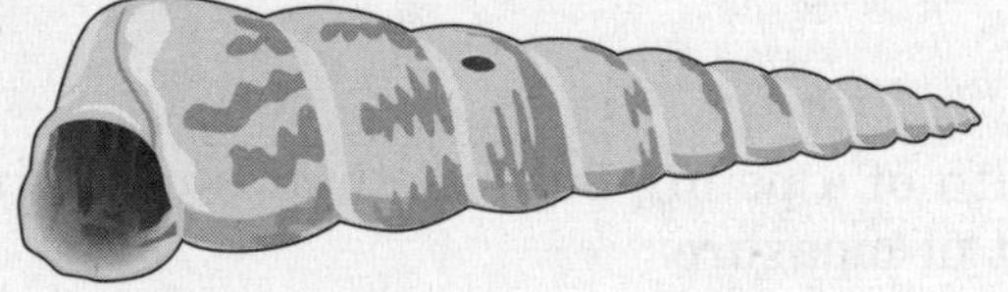

10.

11.

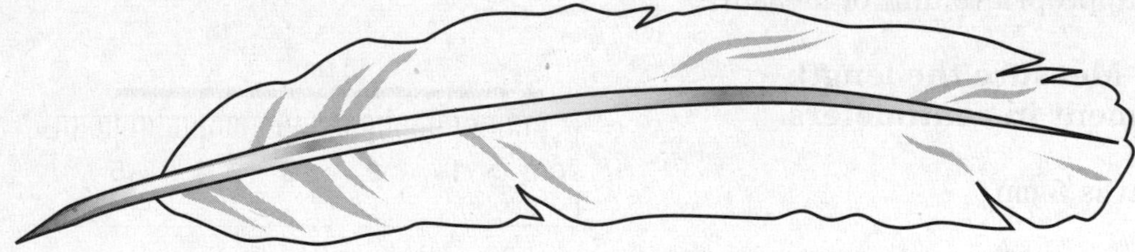

12.

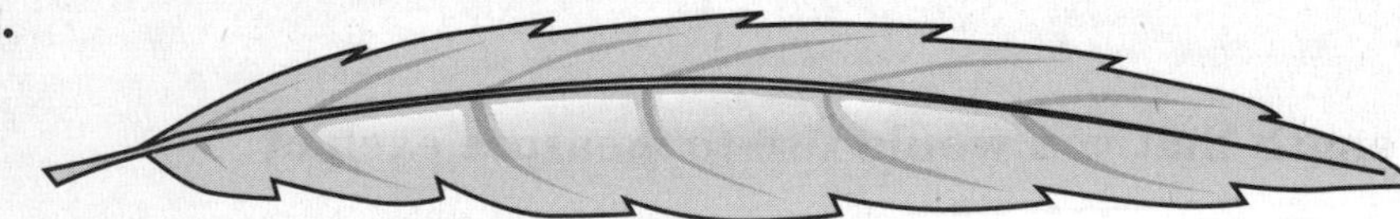

13. Which is greater: 6,200 meters or 5 kilometers? Explain your reasoning.

14. Which is less: 2 kilometers or 1 mile? Explain your reasoning.

15. **PICTURE FRAMES** Yolanda is making a square picture frame from four pieces of wood. Should she be accurate to the nearest meter, to the nearest centimeter, or to the nearest millimeter? Explain your reasoning.

Copyright © Glencoe/McGraw-Hill, a division of The McGraw-Hill Companies, Inc.

8-4 Study Guide and Intervention

Mass and Capacity in the Metric System

In the metric system, the most commonly used units of mass are the milligram (mg), gram (g), and kilogram (kg). One milligram is 0.001 gram and 1 kilogram is 1,000 grams.

Metric Units of Mass		
Unit	**Model**	**Benchmark**
1 **milligram** (mg)	grain of salt	1 mg ≈ 0.00004 oz
1 **gram** (g)	small paper clip	1 g ≈ 0.04 oz
1 **kilogram** (kg)	six medium apples	1 kg ≈ 2 lb

Example 1 **Write the metric unit of mass that you would use to measure a portable radio. Then estimate the mass.**

A portable radio has a mass greater than six apples. So, the kilogram is the appropriate unit.

Estimate Since a portable radio is about three times heavier than six apples, the mass of a portable radio is about 3 kilograms.

The basic unit of capacity in the metric system is the liter (L). One milliliter (mL) is 0.001 liter.

Metric Units of Capacity		
Unit	**Model**	**Benchmark**
1 **milliliter** (mL)	eyedropper	1 mL ≈ 0.03 fl oz
1 **liter** (L)	small pitcher	1 L ≈ 1 qt

Example 2 **Write the metric unit of capacity that you would use to measure a mug of soup. Then estimate the capacity.**

A mug of soup is greater than an eyedropper and less than a small pitcher. So, the milliliter is the appropriate unit.

Estimate There are 1,000 milliliters in a liter. A small pitcher can fill about 4 mugs. So, a mug of soup is about 1,000 ÷ 4 or about 250 milliliters.

Exercises

Write the metric unit of mass or capacity that you would use to measure each of the following. Then estimate the mass or capacity.

1. peanut

2. serving of salad dressing

3. eyelash

4. large soda bottle

5. bottle of milk

6. house cat

7. screw

8. pencil

Copyright © Glencoe/McGraw-Hill, a division of The McGraw-Hill Companies, Inc.

NAME _______________________ DATE __________ PERIOD _____

8-4 Practice

Mass and Capacity in the Metric System

Write the metric unit of mass or capacity that you would use to measure each of the following. Then estimate the mass or capacity.

1. ballpoint pen

2. grain of sand

3. bear

4. egg

5. nickel

6. bowling ball

7. small feather

8. liquid in a thermometer

9. shampoo bottle

10. plastic wading pool

11. hummingbird feeder

12. banana

ANALYZE TABLES **For Exercises 13 and 14, use the table at the right that shows the mass of squirrels.**

Squirrel Masses	
Squirrel	**Average Mass (g)**
African Pygmy	10
Eastern Gray	553
Red	285
Thirteen-Lined Ground	543

Sources: *Animal Fact File* and *National Audubon Society First Field Guide*

13. Is the combined mass of the African Pygmy squirrel, the Eastern Gray squirrel, and the Red squirrel more or less than one kilogram?

14. Which squirrels from the table will have a combined mass closest to one kilogram? Explain your reasoning.

15. COOKING Cooking oil comes in 1.42 liter bottles and 710 milliliter bottles. Which bottle is larger? Explain.

16. BEVERAGES A kiloliter is equal to 1,000 liters and is about the amount needed to fill 5 bathtubs. Each year, about 198 liters of soda is consumed per person in the United States. About how many bathtubs could be filled with the amount of soda drunk by 15 persons in a year?

17. MEDICINE A multivitamin tablet contains 162 milligrams of calcium. If you take one vitamin tablet each day, how many milligrams of calcium will you consume in a week?

Copyright © Glencoe/McGraw-Hill, a division of The McGraw-Hill Companies, Inc.

NAME ________________________________ DATE ____________ PERIOD ______

8-5 Study Guide and Intervention

Problem-Solving Investigation: Use Benchmarks

When solving problems, one strategy that is helpful is to *use benchmarks.* A benchmark is a measurement by which other items can be measured. Sometimes you might not have the exact measuring tool to solve a problem. When this happens, you can use a benchmark to solve the problem.

You can use the *benchmark* strategy, along with the following four-step problem solving plan to solve a problem.

1 Understand – Read and get a general understanding of the problem.

2 Plan – Make a plan to solve the problem and estimate the solution.

3 Solve – Use your plan to solve the problem.

4 Check – Check the reasonableness of your solution.

Example

CRAFTS **Aaliyah is buying fabric to make curtains for her bedroom. She needs 20 feet of fabric to make the curtains. Aaliyah knows that the distance from her nose to her finger when her arm is stretched out is about 1 yard. How can Aaliyah make sure she buys enough fabric without a measuring device?**

Understand Aaliyah needs to measure enough fabric for 20 feet.

Plan She can use the estimated measure from her nose to her finger to measure out enough yards of fabric to be at least 20 feet.

Solve Find how many yards of fabric she needs.

$20 \text{ ft} = \underline{\ ?\ } \text{ yd}$

$20 \text{ ft} \div 3 = 6\frac{2}{3} \text{ yd}$

Aaliyah should count and measure the distance from her nose to her finger 7 times so that she has enough fabric.

Check Since 7 yards is equal to 21 feet, Aaliyah should have enough fabric using her estimated measurement.

Exercise

PETS At a dog show, Brandon was asked the length of his dog. He did not have a measuring device handy, but knew that the width of his hand was about 1 decimeter. Describe a way Brandon could estimate the length of his dog in centimeters.

Copyright © Glencoe/McGraw-Hill, a division of The McGraw-Hill Companies, Inc.

NAME ______________________________ DATE ____________ PERIOD _____

8-5 Practice

Problem-Solving Investigation: Use Benchmarks

Mixed Problem Solving

Use the benchmark strategy to solve Exercises 1 and 2.

1. **FENCES** Mr. Badilla is building a rectangular fence around his back yard. He needs to buy enough fencing material to cover two lengths and one width. Mr. Badilla knows that his walking stride is about one half meter long. Describe a way Mr. Badilla could estimate the amount of fencing material he will need.

2. **BEDROOM** Mindy and her sister share a bedroom. They want to divide the room into a separate space for each of them by putting up a curtain of bed sheets. Sheets are 2 meters wide. Mindy has string and knows that the width of the bedroom door is 1 meter. Describe a way Mindy can estimate the number of sheets to buy.

Use any strategy to solve Exercises 3–5. Some strategies are shown below.

Problem-Solving Strategies
• Guess and check.
• Look for a pattern.
• Use benchmarks.

3. **MUSEUM** The table shows attendance at a museum during the past months. Which is greater, the mean or the median attendance during this time?

Museum Attendance
840 900 725 600 700 985 625
960 825 800 841 900 725

4. **PICTURES** Jon is putting a ribbon border around some picture frames. He knows that the length of his smallest finger is 6 centimeters. Describe a way that Jon can determine how much ribbon he will need.

5. **MONEY** Theo bought a coat that sold for \$108.59. He paid a total of \$115.11, which included tax. How much did he pay in tax?

Copyright © Glencoe/McGraw-Hill, a division of The McGraw-Hill Companies, Inc.

NAME ______________________ DATE ____________ PERIOD ____

8-6 Study Guide and Intervention

Changing Metric Units

To change from one unit to another within the metric system, you can either multiply or divide by powers of ten.

1,000	100	10	1	0.1	0.01	0.001
thousands	hundreds	tens	ones	tenths	hundredths	thousandths
kilo	hecto	deka	basic unit	deci	centi	milli

Each place value is 10 times the place value to its right.

- To change from larger units to smaller units, multiply.
- To change from smaller units to larger units, divide.

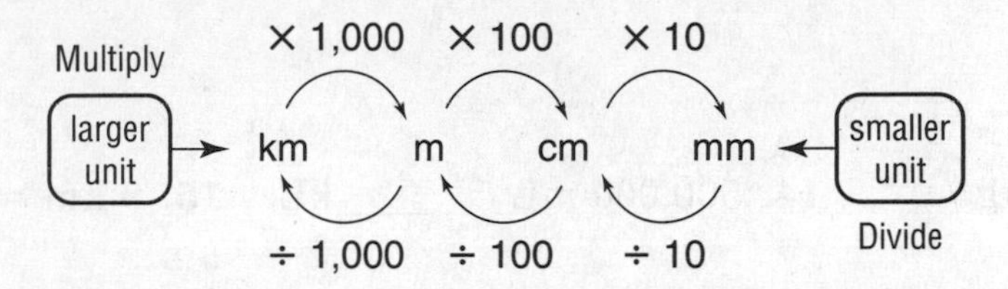

Multiply: larger unit → km (× 1,000) m (× 100) cm (× 10) mm

Divide: smaller unit → mm (÷ 10) cm (÷ 100) m (÷ 1,000) km

Complete.

1 **650 cm = __?__ m**

Since 1 meter = 100 centimeters, divide by 100. 650 ÷ 100 = 6.5
So, 650 cm = 6.5 m.

2 **__?__ mL = 3 L**

Since 1 liter = 1,000 milliliters, multiply by 1,000. 3 × 1,000 = 3,000
So, 3,000 mL = 3 L.

3 **9,100 g = __?__ kg**

Since 1 kilogram = 1,000 grams, divide by 1,000. 9,100 ÷ 1,000 = 9.1
So, 9,100 g = 9.1 kg.

CHECK Since a kilogram is a larger unit than a gram, the number of kilograms should be less than the of grams. The answer seems reasonable.

Exercises

Complete.

1. 2 L = __?__ mL
2. 400 mm = __?__ cm
3. 8 g = __?__ mg

4. 25 cm = __?__ mm
5. 4,100 cm = __?__ m
6. 3 m = __?__ mm

7. __?__ m = 5 km
8. 1,900 g = __?__ kg
9. 6 kg = __?__ g

10. 62 L = __?__ mL
11. 900 mg = __?__ g
12. __?__ km = 500 m

Lesson 8–6

Copyright © Glencoe/McGraw-Hill, a division of The McGraw-Hill Companies, Inc.

NAME ______________________ DATE ____________ PERIOD _____

8-6 Practice

Changing Metric Units

Complete.

1. 91 mm = _____ cm

2. 2 m = _____ mm

3. _____ L = 12 mL

4. _____ mg = 8 g

5. _____ g = 2,500 mg

6. _____ mL = 572 L

7. 21 L = _____ mL

8. 432 cm = _____ m

9. _____ L = 821 mL

10. 2,900 g = ___ kg

11. 670 m = _____ km

12. _____ g = 3 mg

13. 300 mg = _____ kg

14. 500,000 mL = _____ kL

15. 9 km = _____ cm

Order each set of measurements from least to greatest.

16. 6.4 kg, 640 g, 600,000 mg

17. 3.4 km, 33 cm, 340 mm

18. **ANIMALS** An ostrich, the world's largest flightless bird, has a mass of 136 kilograms. A bee hummingbird, the world's smallest bird, has a mass of 2 grams. How much more mass does the ostrich have than the hummingbird?

19. **TRAIL** Hal is hiking a trail to a waterfall viewing platform. It is 8,000 meters to the river, then 10,000 meters from the river to the base of the waterfall, and 600 meters further to the viewing platform. How many kilometers will Hal hike to reach the viewing platform?

Copyright © Glencoe/McGraw-Hill, a division of The McGraw-Hill Companies, Inc.

NAME ______________________ DATE ____________ PERIOD _____

8-7 Study Guide and Intervention

Measures of Time

Units of Time	
Unit	**Model**
1 **second** (s)	time needed to say 1,001
1 **minute** (min) = 60 seconds	time for 2 average TV commercials
1 **hour** (h) = 60 minutes	time for 2 weekly TV sitcoms

To add or subtract measures of time, first add or subtract the seconds, next add or subtract the minutes, and then add or subtract the hours. Rename if necessary in each step.

Example 1 **Find the sum of 3 h 14 min 12 s and 4 h 48 min 3 s.**

Estimate 3 h 14 min 12 s is about 3 h, and 4 h 48 min 3 s is about 5 h.
3 h + 5 h = 8 h.

```
  3 h 14 min 12 s
+ 4 h 48 min  3 s
  7 h 62 min 15 s
```

Add seconds first, then minutes, and finally hours. 62 minutes equals 1 hour 2 minutes.

So, the sum is 8 h 2 min 15 s.
Compare the answer to the estimate.

Example 2 **Find the difference of 5 h 7 min 20 s and 2 h 25 min 12 s.**

Estimate 5 h 7 min 20 s is about 5 h, and 2 h 25 min 12 s is about 2 h.
5 h − 2 h = 3 h.

```
  5 h  7 min 20 s
− 2 h 25 min 12 s
```

Since you cannot subtract 25 minutes from 7 minutes, you must rename 5 hours 7 minutes as 4 hours 67 minutes.

```
  5 h  7 min 20 s          4 h 67 min 20 s
− 2 h 25 min 12 s   →    − 2 h 25 min 12 s
                           2 h 42 min  8 s
```

Compare the answer to the estimate.

Example 3 **Gloria practiced on her flute from 11:40 A.M. until 1:52 P.M. How long did she practice?**

You need to find how much time has elapsed. Gloria's practice time is
20 minutes + 1 hour 52 minutes or
1 hour 72 minutes. Now rename
72 minutes as 1 hour 12 minutes.
1 h + 1 h 12 min = 2 h 12 min.
Gloria practiced for 2 hours 12 minutes.

11:40 A.M. to 12:00 noon is 20 minutes.

12:00 noon to 1:52 P.M. is 1 hour 52 minutes.

Exercises

Add or subtract.

1.
```
  4 h 18 min 11 s
− 3 h 15 min  4 s
```

2.
```
  6 h  7 min 42 s
+ 2 h 12 min 38 s
```

3.
```
  5 h 18 min 12 s
− 2 h  6 min 41 s
```

Find the elapsed time.

4. 4:25 P.M. to 11:55 P.M.

5. 9:20 A.M. to 5:05 P.M.

6. 10:30 A.M. to 1:43 P.M.

Lesson 8–7

Copyright © Glencoe/McGraw-Hill, a division of The McGraw-Hill Companies, Inc.

NAME ______________________ DATE ____________ PERIOD _____

8-7 Practice

Measures of Time

Add or subtract.

1. 23 h 52 min
+ 15 h 30 min

2. 17 min 14 s
+ 32 min 50 s

3. 8 h 39 min 43 s
− 3 h 57 min 22 s

4. 4 h 49 min 28 s
− 2 h 12 min 53 s

5. 28 h 13 min 26 s
+ 3 h 58 min 36 s

6. 9 h 37 s
+ 2 h 6 min 50 s

7. 4 h 39 s
5 h 21 min 8 s
+ 49 min 11 s

8. 5 h 47 s
+ 52 s

9. 6 h
− 2 h 35 min 41 s

Find the elapsed time.

10. 8:30 A.M. to 11:43 A.M.

11. 4:32 P.M. to 8:12 P.M.

12. 5:45 A.M. to 3:32 P.M.

13. 8:30 P.M. to 10:53 A.M.

14. **ACTIVITIES** On Saturday, Miguel spent 2 hours 30 minutes at basketball practice and 90 minutes at his pastry chef class. How much total time did Miguel spend doing these two activities?

15. **FURNITURE** Una is painting a chair. She finished painting the first coat at 10:42 A.M. The paint needs to dry for at least 1 hour 45 minutes before another coat of paint can be put on. At what time will Una be able to paint a second coat?

Copyright © Glencoe/McGraw-Hill, a division of The McGraw-Hill Companies, Inc.

8-8 Study Guide and Intervention

Measures of Temperature

Temperature is the measure of hotness or coldness of an object or environment. Temperature is measured in **degrees**.

Temperature in the metric system is measured in degrees **Celsius (°C)**. Temperature in the customary system is measured in degrees **Fahrenheit (°F)**.

Example 1 **Which is a more reasonable temperature for a glass of orange juice, 10ºC or −20ºC?**

Water freezes at 0ºC, so −20ºC is below freezing. A more reasonable temperature of a glass of orange juice is 10ºC.

Example 2 **What is a reasonable estimate of the temperature in degrees Celsius and degrees Fahrenheit on a fall day?**

A fall day would have a temperature between a warm summer temperature of 90ºF and a cold winter temperature of 30ºF.

So, a reasonable fall temperature would be 60ºF or 15ºC.

Exercises

Choose the more reasonable temperature for each.

1. a newborn baby's temperature: 99ºF or 129ºF

2. pie in oven: 75ºC or 175ºC

3. ice cream sandwich: −15ºF or 15ºF

4. cup of hot chocolate: 140ºF or 240ºF

Give a reasonable estimate of the temperature in degrees Celsius and degrees Fahrenheit for each activity.

5. inside a basement

6. planting a garden outside

7. water skiing

8. temperature in a freezer

Copyright © Glencoe/McGraw-Hill, a division of The McGraw-Hill Companies, Inc.

Lesson 8–8

NAME ______________________ DATE ____________ PERIOD _____

8-8 Practice

Measures of Temperature

Choose the more reasonable temperature for each.

1. inside your bedroom: 68°F or 120°F

2. chocolate chips starting to melt in your hand: 60°F or 100°F

3. snowflake: −15°C or 10°C

4. lava flowing from a volcano: 70°C or 100°C

5. rain in a tropical forest: 10°C or 40°C

6. water in a goldfish bowl: 68°F or 100°F

Give a reasonable estimate of the temperature in degrees Celsius and degrees Fahrenheit for each activity.

7. surfing the Internet with a computer

8. water skiing

9. sitting in front of a fire burning in a fireplace

10. playing table tennis

11. sledding down a snow bank

12. playing football

13. **AIR CONDITIONER** The Johnsons purchased an air conditioner. Should they set the thermostat at 18°C or 50°C? Explain your reasoning.

14. **AUTUMN** It is a cool, crisp autumn day. If the temperature reads 15 degrees, is this 15°C or 15°F?

15. **PAINTING** A paint can says not to paint if the air temperature is above 80°F. Your thermometer says the air temperature is 35°C. Should you paint today? Explain your reasoning.

16. **CHEMISTRY** Iron is heated in the process of making steel. Iron starts to turn from a solid into a liquid at about 5,182°F. About how much hotter must iron be to start melting than frozen water needs to be to start melting?

Copyright © Glencoe/McGraw-Hill, a division of The McGraw-Hill Companies, Inc.

NAME ______________________ DATE ____________ PERIOD _____

9-1 Study Guide and Intervention

Measuring Angles

Angles have two **sides** that share a common endpoint called the **vertex**.

Angles are measured in **degrees**. One degree is equal to $\frac{1}{360}$th of a circle.

Angles can be classified according to their measure.

vertex → sides

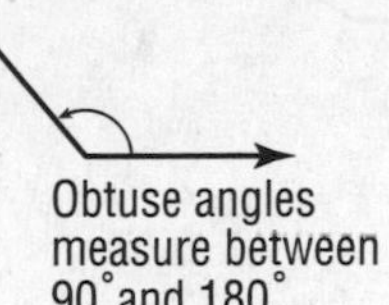

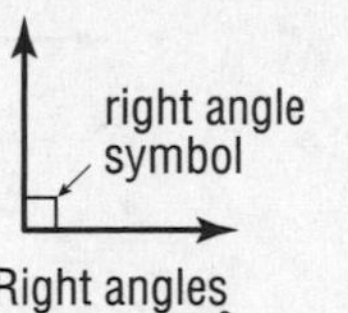

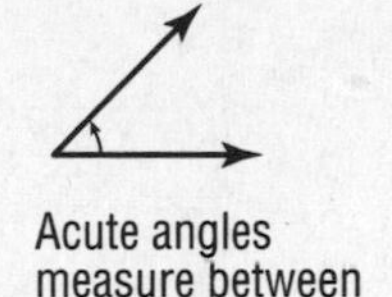

Example 1 **Use a protractor to find the measure of the angle.**

To measure an angle, place the center of a protractor on the vertex of the angle. Place the zero mark of the scale along one side of the angle. Then read the angle measure where the other side of the angle crosses the scale.

The angle measures 30°.

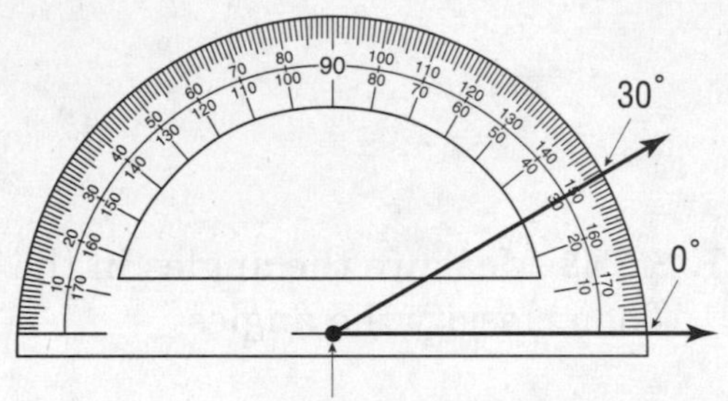

Align the center of the protractor. This angle measures 30°.

Example 2 **Classify the angle at the right as *acute, obtuse, right,* or *straight.***

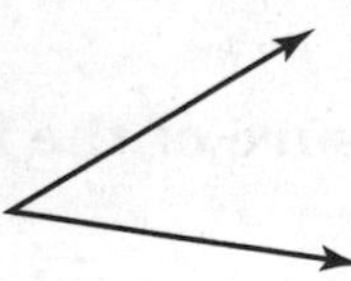

The angle smaller than a right angle, so it is less than 90°

The angle is an acute angle.

Exercises

Use a protractor to find the measure of each angle. Then classify each angle as *acute, obtuse, right,* or *straight.*

1. 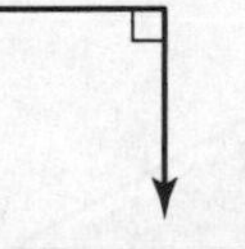

2. 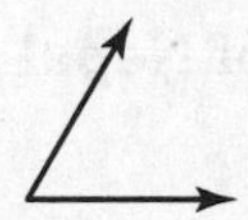

3. 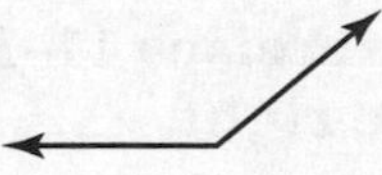

4. ←——•——→

5. 

6. 

Lesson 9–1

Copyright © Glencoe/McGraw-Hill, a division of The McGraw-Hill Companies, Inc.

NAME ______________________ DATE __________ PERIOD _____

9-1 Practice

Measuring Angles

Use a protractor to find the measure of each angle. Then classify each angle as *acute*, *obtuse*, *right*, or *straight*.

1. 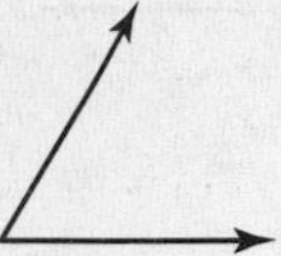

2. 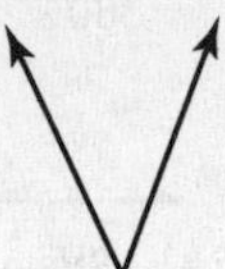

3. 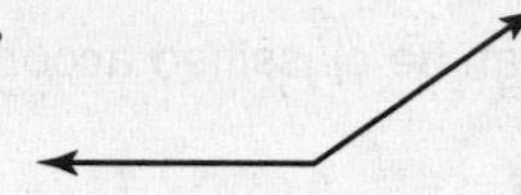

4. 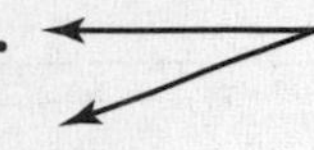

5. 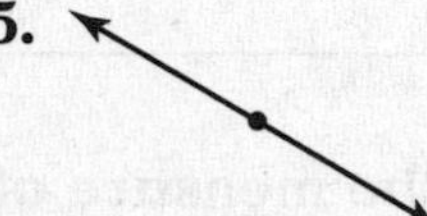

6. 

7. **SIGNS** Measure the angles in the road sign. Then classify the angles.

Find the measure of the indicated angle in each figure.

8. 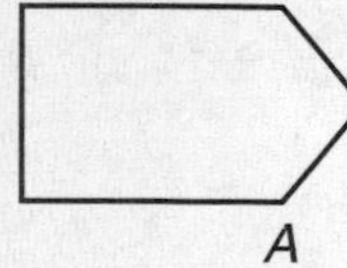

9. 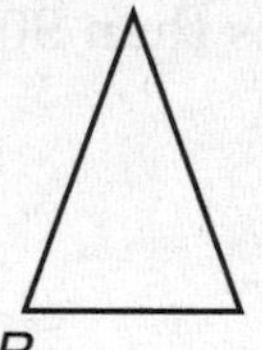

10. 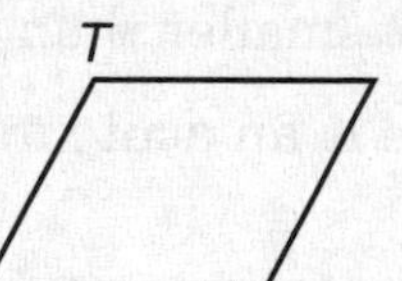

FLAGS For Exercises 11–13, refer to the flag of Nepal shown at the right.

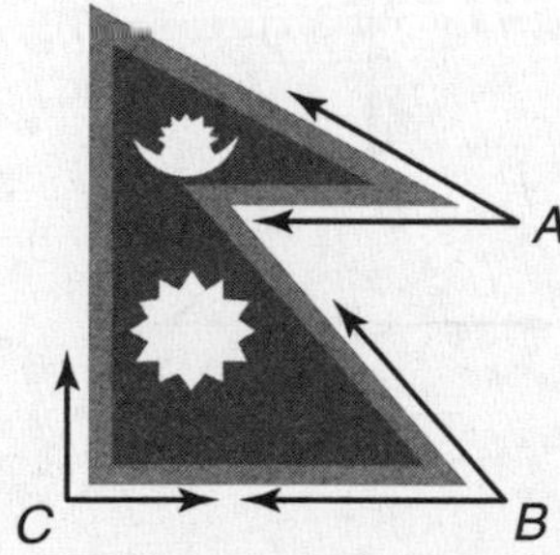

11. What is the measure of $\angle A$?

12. What is the measure of $\angle B$?

13. Classify $\angle A$, $\angle B$, and $\angle C$.

Copyright © Glencoe/McGraw-Hill, a division of The McGraw-Hill Companies, Inc.

NAME ______________________ DATE ____________ PERIOD _____

9-2 Study Guide and Intervention

Estimating and Drawing Angles

To estimate the measure of an angle, compare it to an angle whose measure you know.

You can use the measures of these angles to estimate measures of other angles.

A protractor and a straightedge can be used to draw angles.

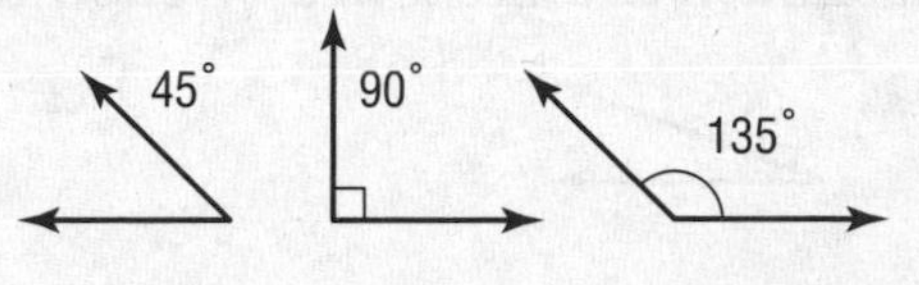

Example 1 **Estimate the measure of the angle shown.**

Compare the given angle to the angles shown above.

The angle is a little greater than 45°, so a reasonable estimate is about 45°.

Example 2 **Draw a 140° angle.**

Step 1
Draw one side. Mark the vertex and draw an arrow.

Step 2
Place the center point of the protractor on the vertex. Align the mark labeled 0 on the protractor with the line. Find 140° and make a dot.

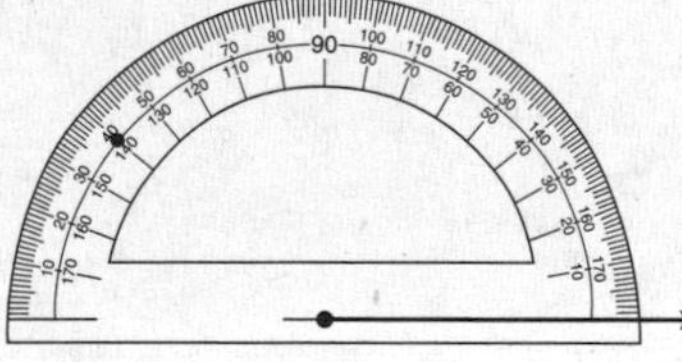

Step 3
Remove the protractor and use the straightedge to draw the side that connects the vertex and the dot.

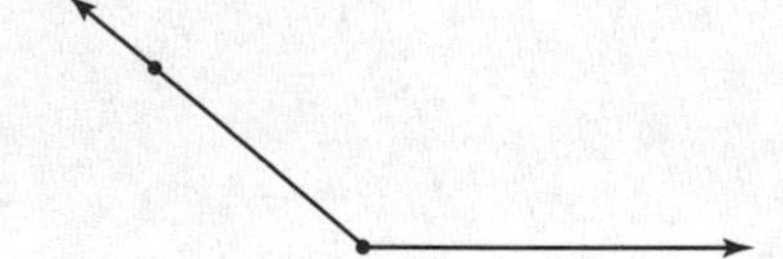

Exercises

Estimate the measure of each angle.

1. 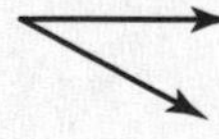

2. 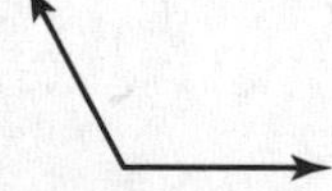

3. 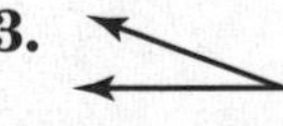

Use a protractor and a straightedge to draw angles having the following measurements.

4. 35°

5. 110°

6. 15°

Copyright © Glencoe/McGraw-Hill, a division of The McGraw-Hill Companies, Inc.

NAME ______________________________ DATE ____________ PERIOD _____

9-2 Practice

Estimating and Drawing Angles

Estimate the measure of each angle.

1. 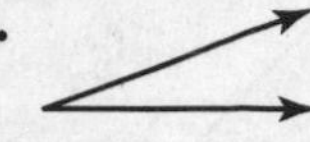

2. 

3. 

4. 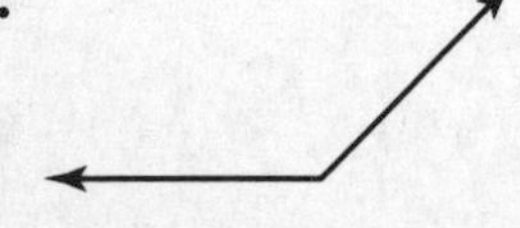

5. 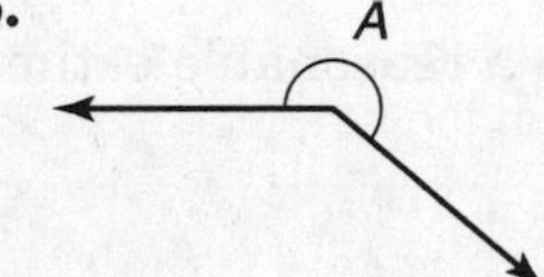

6. 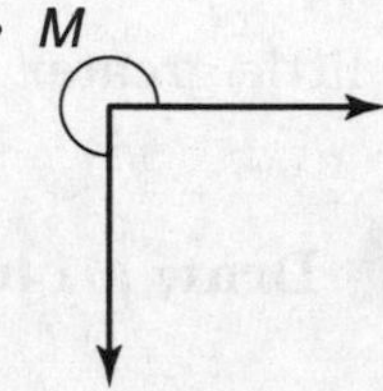

Use a protractor and a straightedge to draw angles having the following measurements.

7. 55°

8. 10°

9. 78°

10. 162°

11. 98°

12. 147°

13. **CASTLES** Caerlaverock Castle in Scotland is built in the shape of a triangle. Each angle of the triangle is 60°. In the space at the right, use a protractor and a straightedge to draw a floorplan of Caerlaverock Castle. Label each angle with its measure.

Copyright © Glencoe/McGraw-Hill, a division of The McGraw-Hill Companies, Inc.

9-3 Study Guide and Intervention

Angle Relationships

Vertical angles are the opposite angles formed by intersecting lines. Vertical angles are **congruent angles**, or angles with the same measure.

Example 1 **Find the value of x in the figure at the right.**

The angle labeled $x°$ and the angle labeled 40° are vertical angles. Therefore, they are congruent.

So, the value of x is 40.

Two angles are **complementary** if the sum of their measures is 90°.
Two angles are **supplementary** if the sum of their measures is 180°

Example 2 **Classify the pair of angles at the right as *complementary, supplementary,* or *neither.***

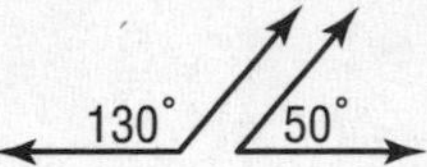

$130° + 50° = 180°$
The angles are supplementary.

Example 3 **Find the value of x in the figure at the right.**

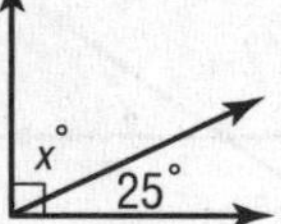

Since the angles form a right angle, they are complementary.

$x° + 25° = 90°$ Definition of complementary angles
$65° + 25° = 90°$ THINK What measure added to 25° equals 90°?

So, the value of x is 65.

Exercises

Classify each pair of angles as *complementary, supplementary,* or *neither.*

1. 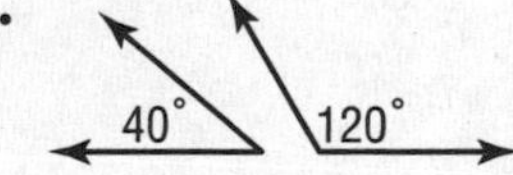

2. 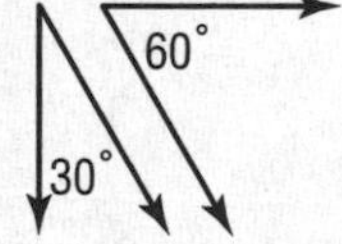

3. 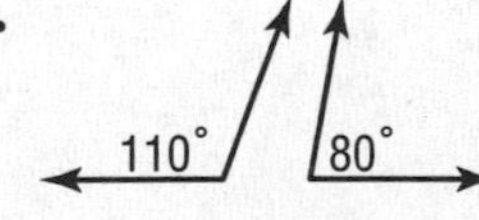

Find the value of x in each figure.

4. 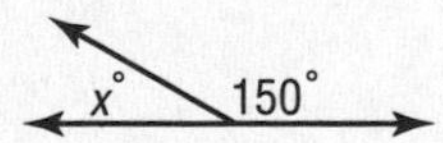

5. 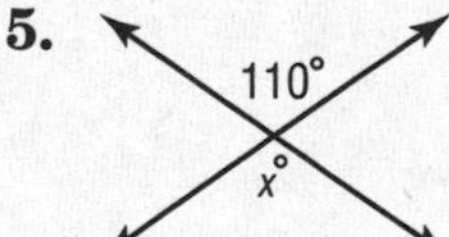

6. 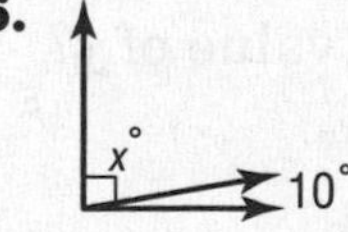

Copyright © Glencoe/McGraw-Hill, a division of The McGraw-Hill Companies, Inc.

9-3 Practice

Angle Relationships

Classify each pair of angles as *complementary, supplementary,* or *neither.*

1.
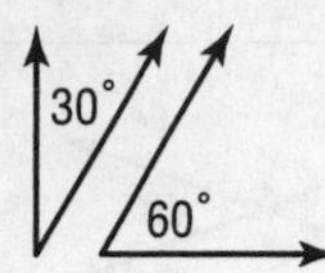

2.
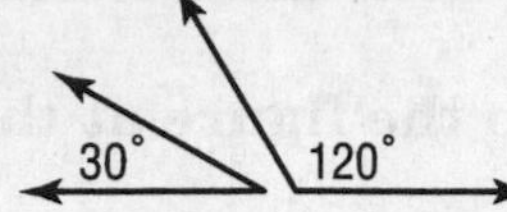

3.
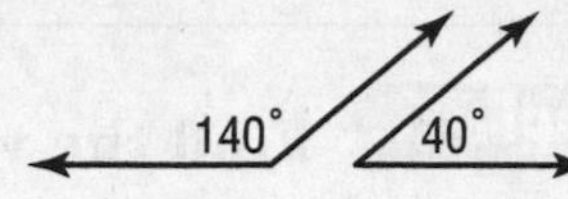

4.
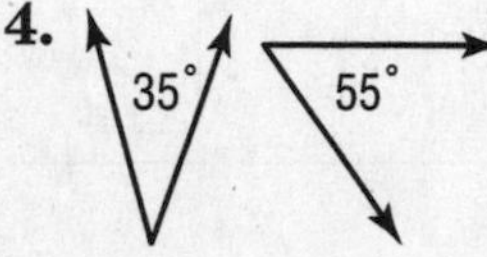

5.
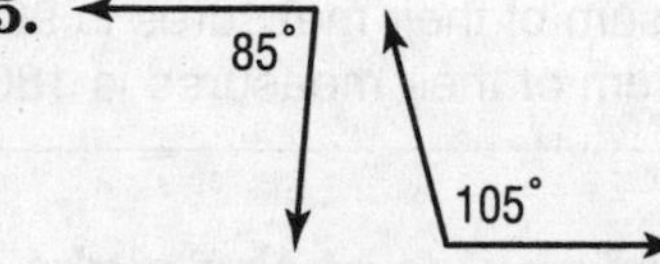

6.
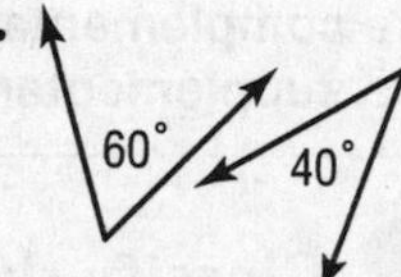

Find the value of *x* in each figure.

7.
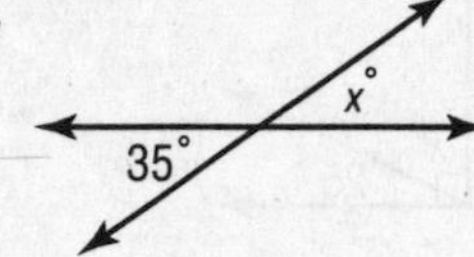

8.
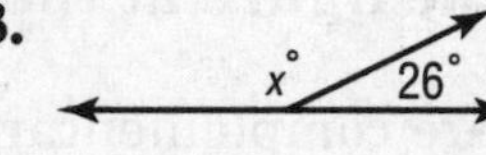

9.
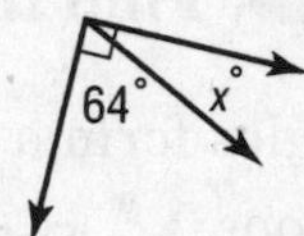

10. Angles A and B are complementary. Find $m\angle B$ if $m\angle A = 71°$.

11. Angles C and D are supplementary. Find $m\angle C$ if $m\angle D = 88°$.

GARDENS **A semicircular garden is divided into four sections as show.**

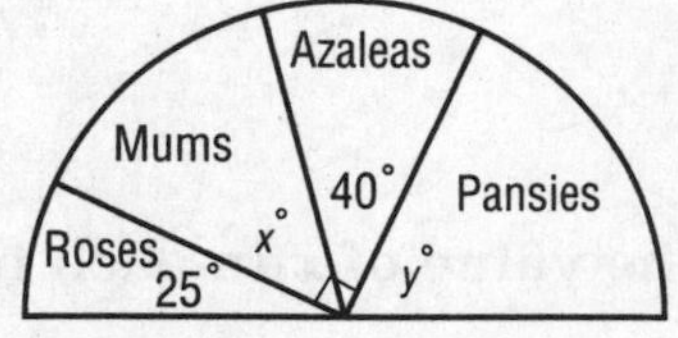

12. What is the value of x?

13. What is the value of y?

Copyright © Glencoe/McGraw-Hill, a division of The McGraw-Hill Companies, Inc.

NAME ______________________________ DATE ____________ PERIOD _____

9-4 Study Guide and Intervention

Triangles

Acute triangles have all acute angles. **Right triangles** have one right angle. **Obtuse triangles** have one obtuse angle. The sum of the angle measures in a triangle is 180°.

Example 1 **Classify the triangle at the right as *acute, right,* or *obtuse*.**

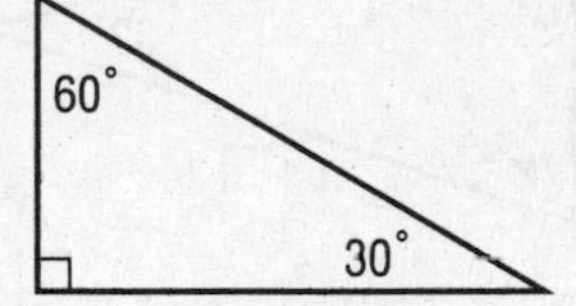

The triangle has one right angle.

So, the triangle is a right triangle.

Example 2 **Find the value of x in the triangle at the right.**

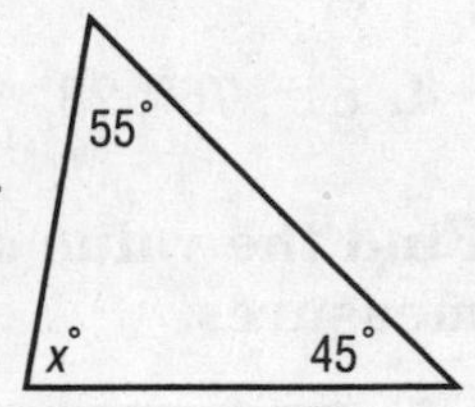

$x° + 55° + 45° = 180°$	The sum of the measures of the angles in a traingle is 180°.
$x° + 100° = 180°$	Add 55° and 45°.
$80° + 100° = 180°$	THINK What measure added to 100° equals 180°?
$x° = 80°$	The solution is 80°.

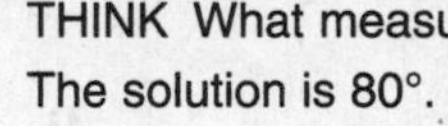

So, the value of x is 80.

Scalene triangles have no congruent sides. **Isosceles triangles** have at least 2 congruent sides. **Equilateral triangles** have 3 congruent sides.

Example 3 **Classify the triangle at the right as *scalene, isosceles,* or *equilateral*.**

8 cm, 8 cm, 5 cm

Two of the sides measure 8 centimeters, and are congruent.

So, the triangle is an isosceles triangle.

Exercises

Classify each triangle as *acute, right,* or *obtuse*.

1. 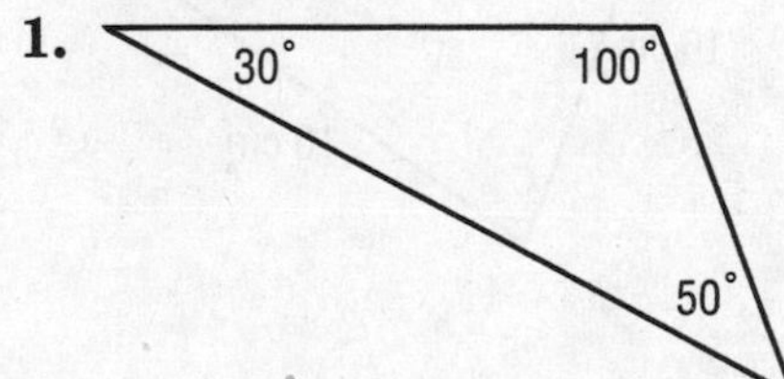

2. 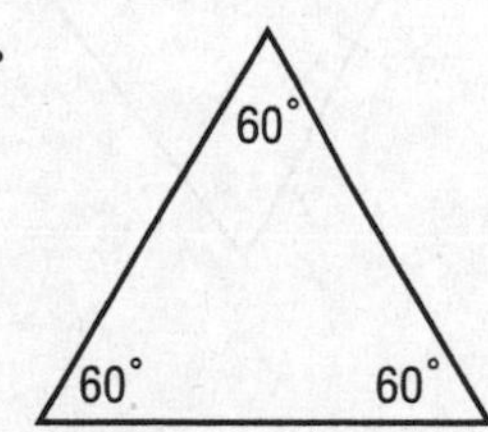

3. 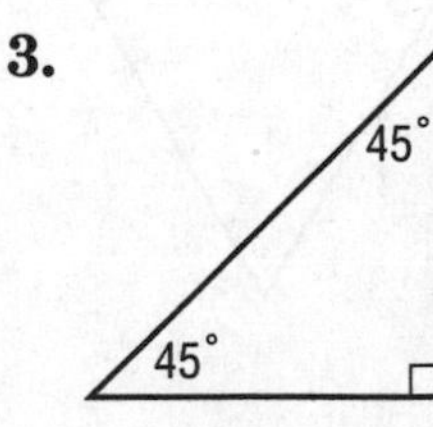

4. Find the value of x in the triangle at the right.

5. Classify the triangle at the right as *scalene*, *isosceles*, or *equilateral*.

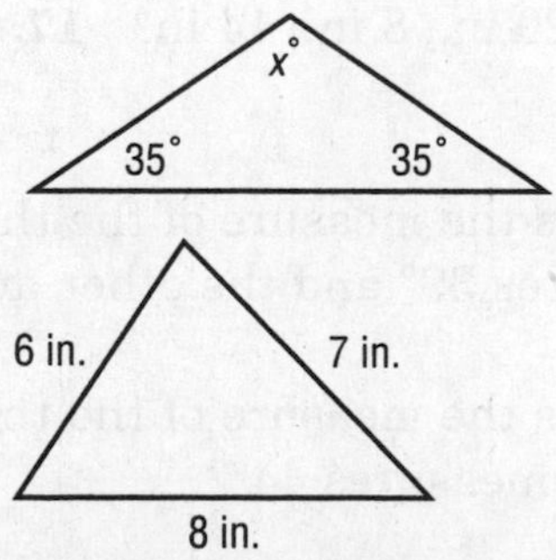

Copyright © Glencoe/McGraw-Hill, a division of The McGraw-Hill Companies, Inc.

NAME ______________________________ DATE ____________ PERIOD _____

9-4 Practice

Triangles

Classify each triangle drawn or having the given angle measures as *acute, right,* or *obtuse.*

1.

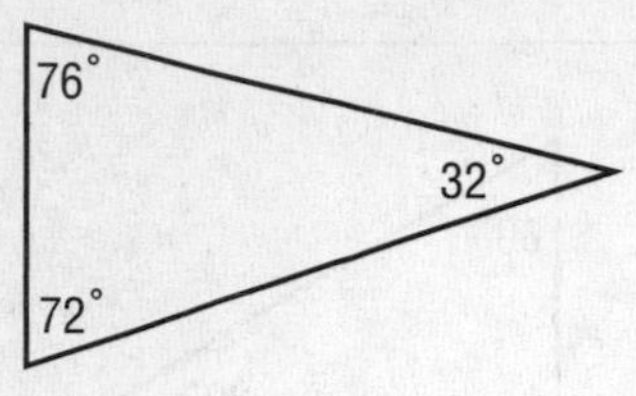

2.

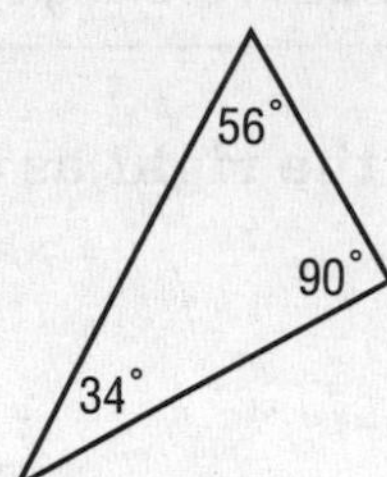

3.

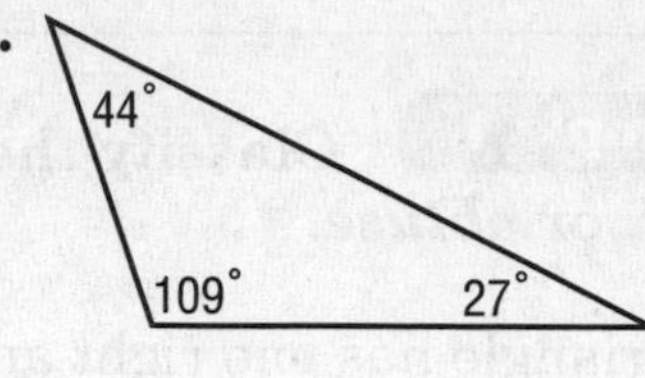

4. 81°, 76°, 23°

5. 118°, 34°, 28°

6. 90°, 60°, 30°

Find the value of *x* in each triangle drawn or having the given angle measures.

7.

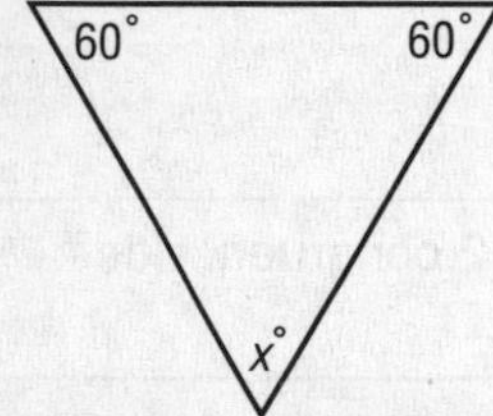

8.

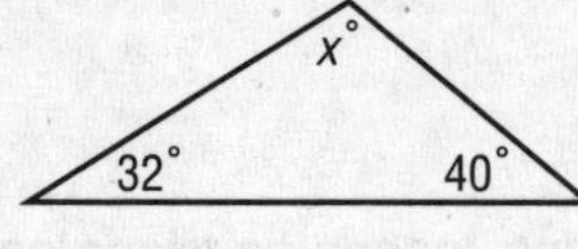

9.

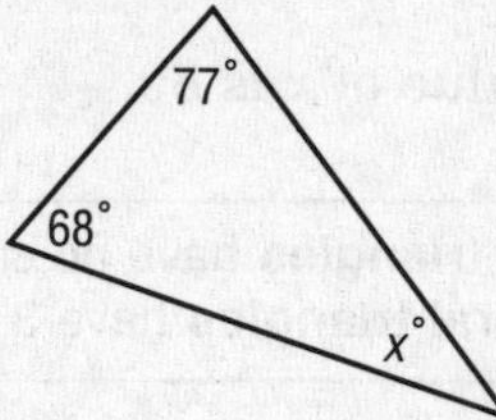

10. 81°, 56°, x°

11. x°, 65°, 21°

12. x°, 42°, 15°

Classify each triangle drawn or described as *scalene, isosceles,* or *equilateral.*

13.

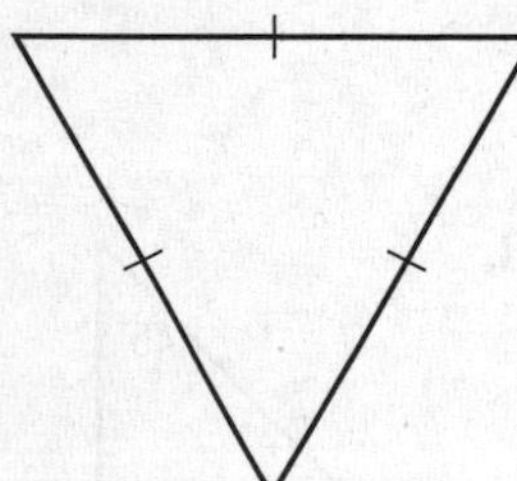

14.

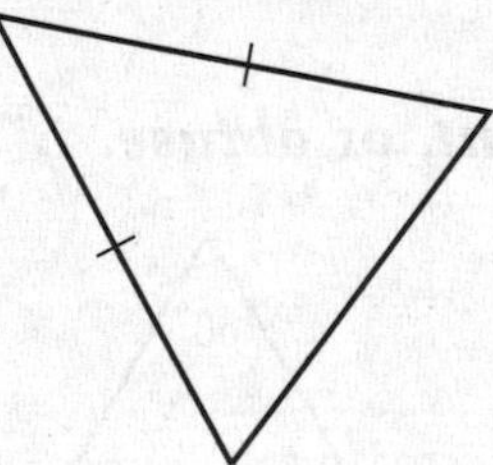

15.

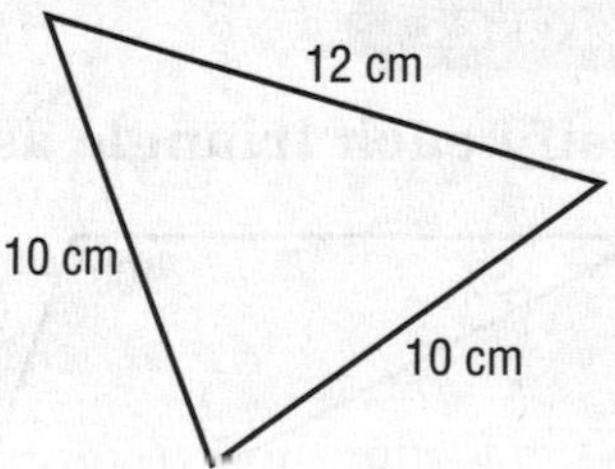

16. sides: 20 in., 8 in., 14 in.

17. sides: 7 ft, 6 ft, 7 ft

18. sides: 4 m, 10 m, 7 m

19. What is the measure of the third angle of a triangle if one angle measures 39° and the other angle measures 78°?

20. What is the measure of the third angle of a right triangle if one of the angles measures 44°?

Copyright © Glencoe/McGraw-Hill, a division of The McGraw-Hill Companies, Inc.

NAME ______________________________ DATE ____________ PERIOD _____

9-5 Study Guide and Intervention

Quadrilaterals

Quadrilaterals have four sides and four angles. The sum of the measures of the angles is 360°.

Example 1 **Find the value of x in the quadrilateral at the right.**

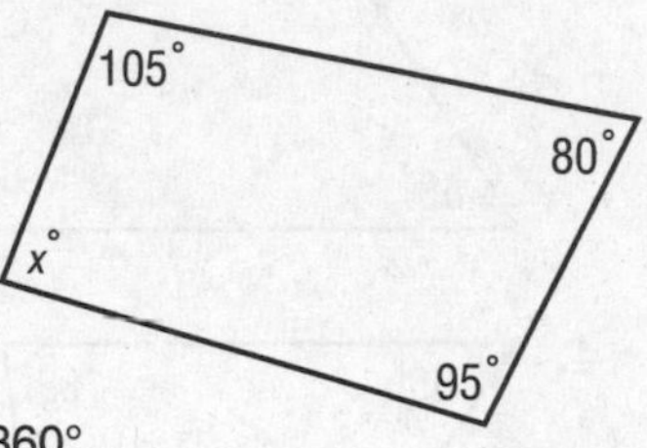

$x° + 105° + 80° + 95° = 360°$ The sum of the measures of the angles of a quadrilateral is 360°.

$x° + 280° = 360°$ Add 105°, 80°, and 95°.

$80° + 280° = 360°$ THINK What measure added to 280° equals 360°.

$x° = 80°$

So, the value of x is 80.

A **rectangle** has opposite sides congruent and parallel, and all right angles.

A **square** has all sides congruent, opposite sides parallel, and all right angles.

A **parallelogram** has opposite sides congruent and parallel, and opposite angles congruent.

A **rhombus** has all sides congruent, opposite sides parallel, and opposite angles congruent.

A **trapezoid** has exactly one pair of opposite sides parallel.

Example 2 **Classify the quadrilateral at the right.**

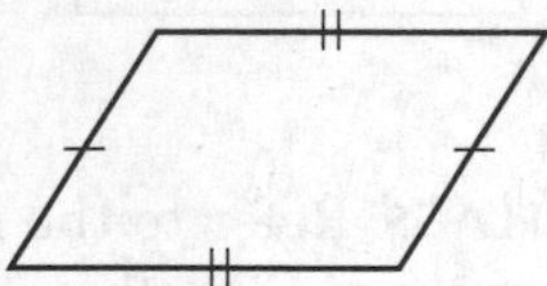

The figure has opposite sides congruent and parallel.

So, the figure is a parallelogram.

Exercises

Find the value of x in each quadrilateral.

1.

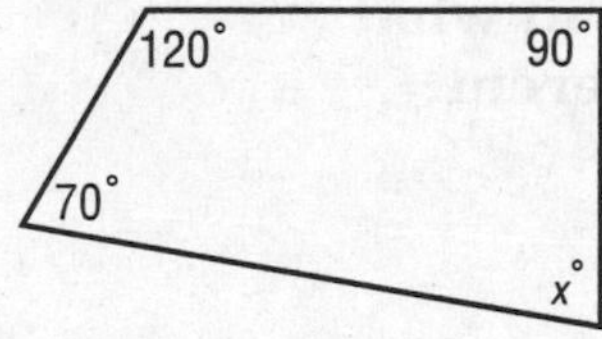

2.

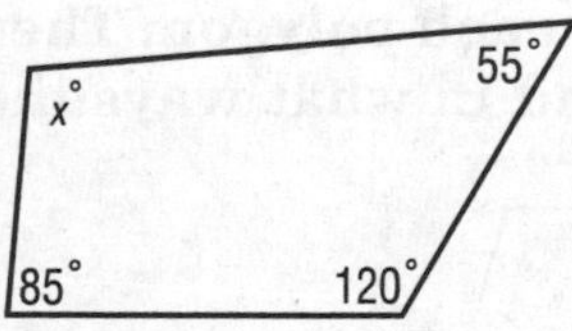

3.

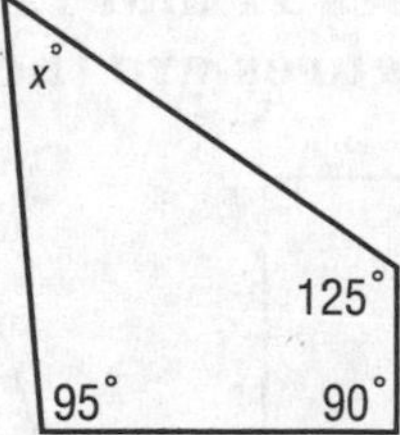

Classify each quadrilateral.

4.

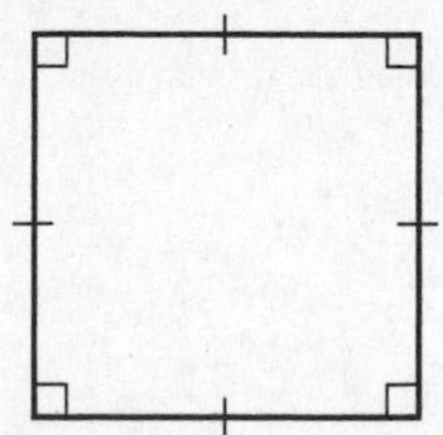

5.

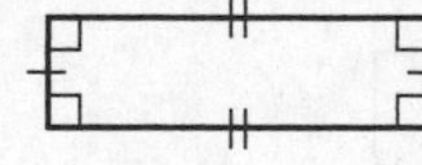

6.

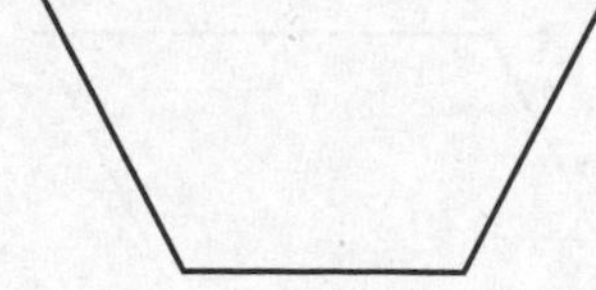

Lesson 9–5

Copyright © Glencoe/McGraw-Hill, a division of The McGraw-Hill Companies, Inc.

NAME ______________________ DATE ____________ PERIOD _____

9-5 Practice

Quadrilaterals

Find the value of *x* in each quadrilateral.

1.
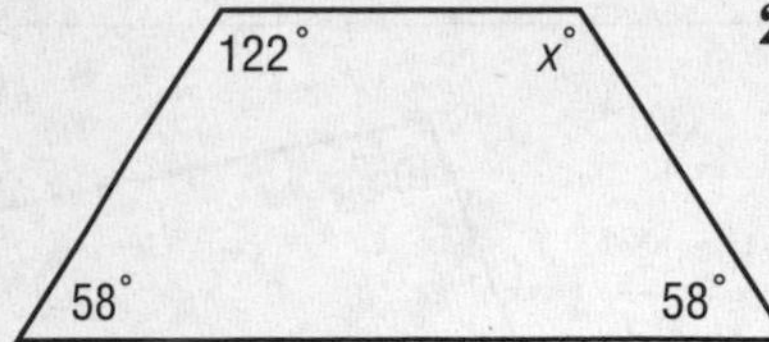

2.
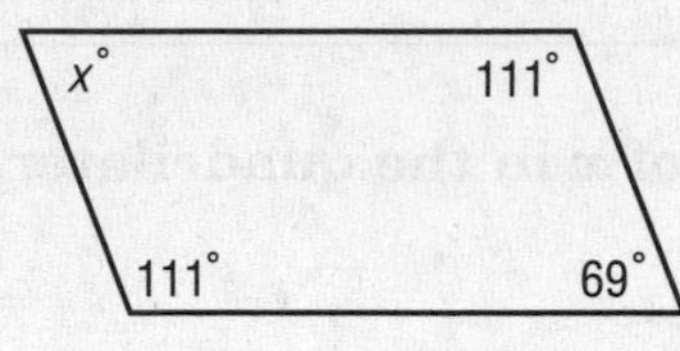

3.
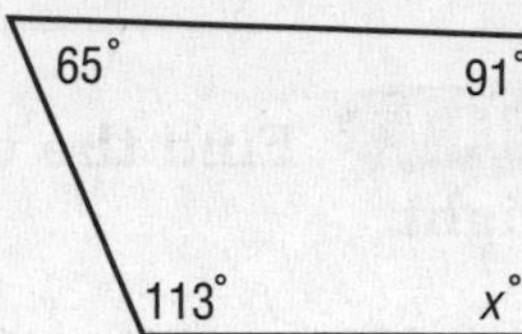

4.
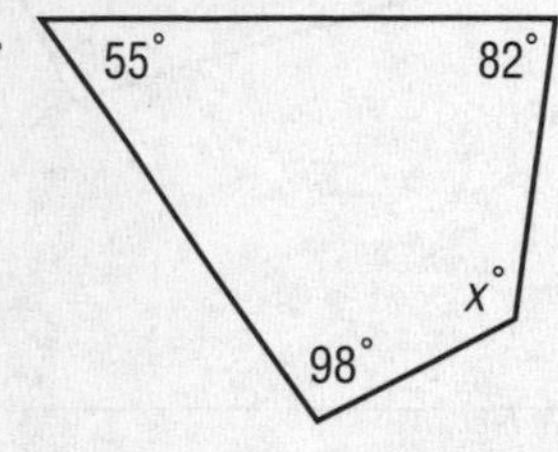

5.
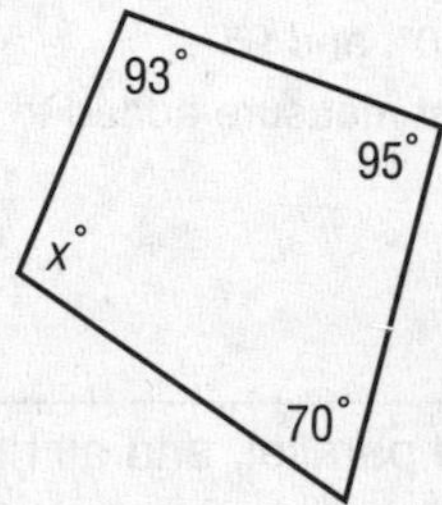

6.
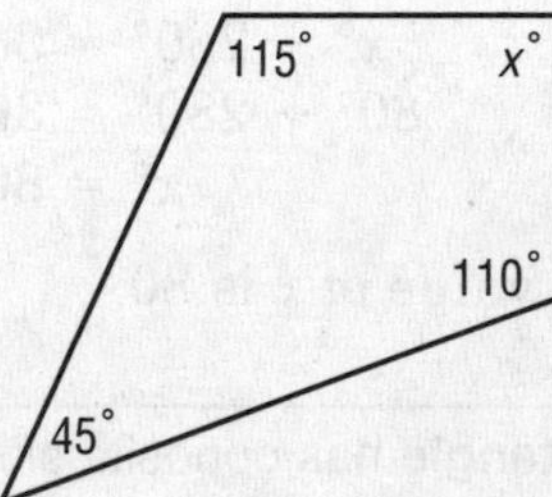

7.
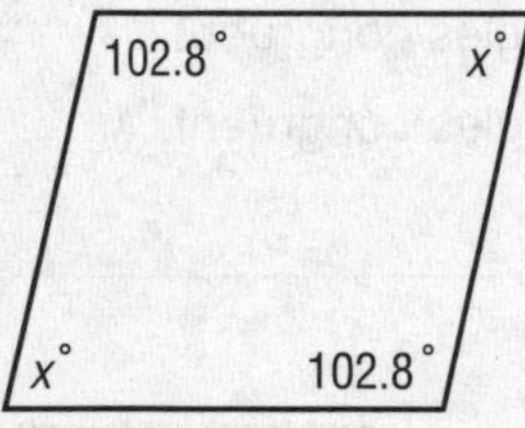

8.
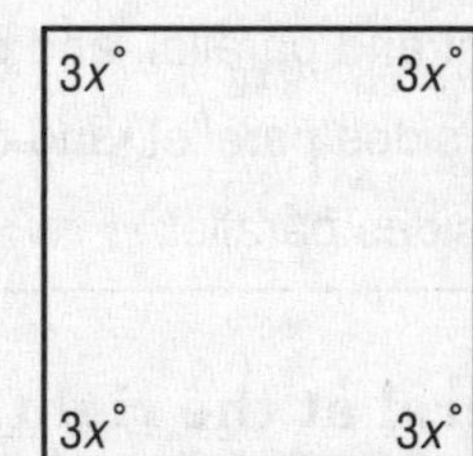

9.
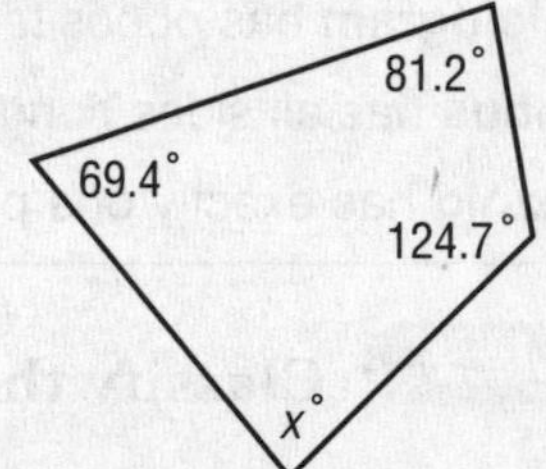

10. **FLAGS** Refer to the flag of Kuwait. Classify the shapes in the flag design.

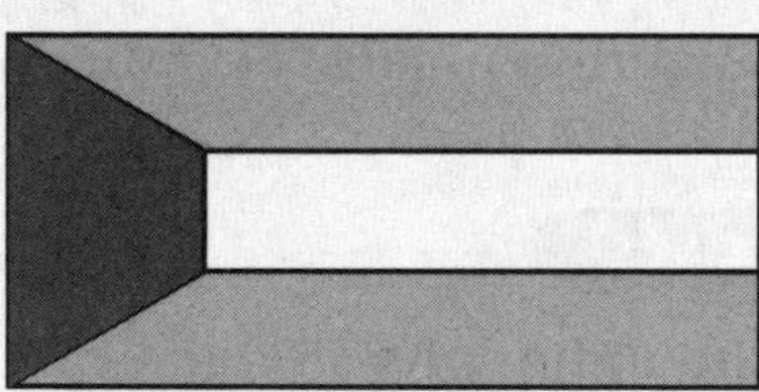

For Exercises 11 and 12, classify each polygon. Then describe in what ways the figures are the same and in what ways they are different.

11.
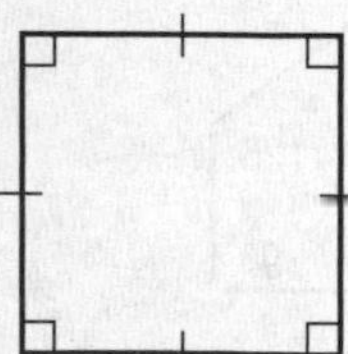

12.
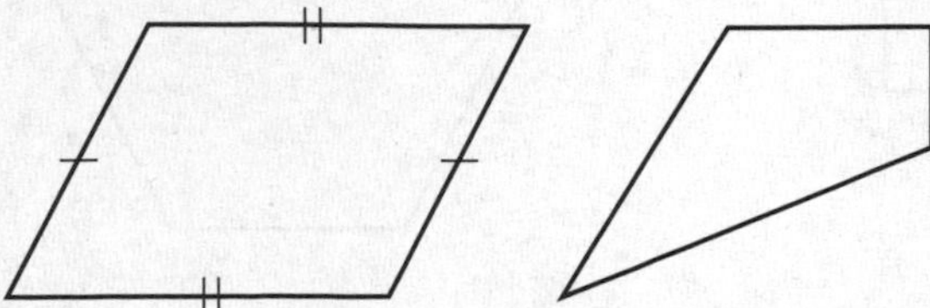

Copyright © Glencoe/McGraw-Hill, a division of The McGraw-Hill Companies, Inc.

NAME ______________________ DATE ____________ PERIOD ____

9-6 Study Guide and Intervention

Problem-Solving Investigation: Draw a Diagram

When solving problems, one strategy that is helpful is to *draw a diagram.* A problem may often describe a situation that is easier to solve visually. You can draw a diagram of the situation, and then use the diagram to solve the problem.

You can draw a diagram, along with the following four-step problem solving plan to solve a problem.

1 Understand – Read and get a general understanding of the problem.

2 Plan – Make a plan to solve the problem and estimate the solution.

3 Solve – Use your plan to solve the problem.

4 Check – Check the reasonableness of your solution.

Example LIBRARY **The school library is putting tables in an open area that is 28 feet by 50 feet. Each table is a square with sides measuring 5 feet, and the tables must be 6 feet apart from each other and the wall. How many tables can fit in this area?**

Understand You know all the dimensions. You need to find how many tables will fit in this area.

Plan Draw a diagram to see how many tables will fit.

Solve

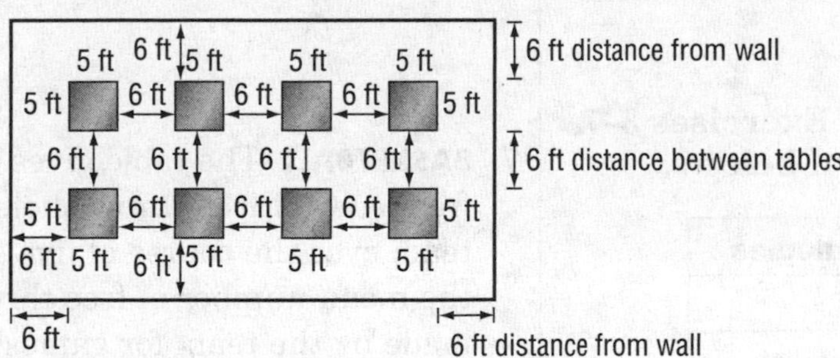

The diagram shows that 8 tables will fit in this area in the library.

Check Make sure the dimensions meet the requirements. The distance across is 50 feet and the distance down is 28 feet. So, the answer is correct.

Exercise

PICTURE FRAME La Tasha is decorating a picture frame by gluing gem stones around the frame. The picture frame is 7 inches by $5\frac{1}{2}$ inches. Each gem stone is $\frac{1}{2}$-inch wide and La Tasha glues them 1 inch apart and 1 inch from the edge. How many gem stones can La Tasha fit on the frame?

Lesson 9–6

Copyright © Glencoe/McGraw-Hill, a division of The McGraw-Hill Companies, Inc.

NAME ______________ DATE ______ PERIOD ______

9-6 Practice

Problem-Solving Investigation: Draw a Diagram

Mixed Problem Solving

Use the draw a diagram strategy to solve Exercises 1 and 2.

1. **RUNNING** Five runners were far ahead in the marathon. Juanita crossed the finish line after Owen and Molly. Molly was first. Juanita was between Greta and Owen. Kenji was last. In what order did the runners cross the finish line?

2. **PLANTS** A nursery is planting seedlings in a plot that is 10 feet by 14 feet. How many seedlings will fit if each seedling is in a 1-foot square peat pot and each peat pot needs to be planted 3 feet apart from another?

Use any strategy to solve Exercises 3–7. Some strategies are shown below.

Problem-Solving Strategies
• Guess and check.
• Make an organized list.
• Look for a pattern.
• Draw a diagram.

3. **PATTERNS** Complete the pattern: 2, 3, 5, 9, __?__, __?__, __?__.

4. **ANIMALS** Jacy is building a fence to create a hexagonal dog pen. Each of the six sides needs four posts. How many posts are needed?

5. **FOOD** A lunch shop offers 2 kinds of soups, 3 kinds of sandwiches, and 3 kinds of beverages. How many combinations of one soup, one sandwich, and one beverage are possible?

6. **GEOMETRY** An official doubles tennis court has a length of 78 feet and a width of 36 feet. How many times greater is the length than the width of the court to the nearest tenth?

7. **BASKETBALL** The table gives the frequency of free throw shots made by a team over the course of five games. Find the mean number of free throw shots made by the team for games 1–5.

Game	Tally	Frequency
1	III	3
2	𝍸	5
3	𝍸 II	7
4	𝍸	5
5	I	1

Copyright © Glencoe/McGraw-Hill, a division of The McGraw-Hill Companies, Inc.

NAME ______________________ DATE ____________ PERIOD _____

9-7 Study Guide and Intervention

Similar and Congruent Figures

Figures that have the same size and shape are **congruent figures**.

Figures that have the same shape but not necessarily the same size are **similar figures**.

Examples **Tell whether each pair of figures is *similar*, *congruent*, or *neither*.**

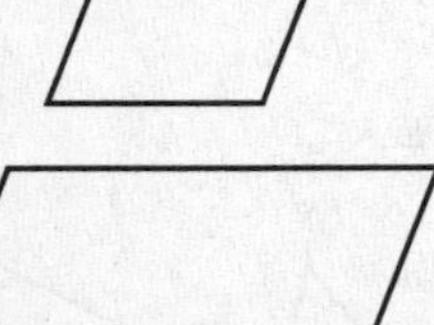

The parallelograms have the same shape but not the same size, so they are similar.

2

The triangles have the same shape and size, so they are congruent.

3

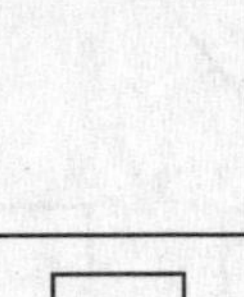

The rectangles are neither the same size nor the same shape, so they are neither congruent nor similar.

Example 4 **The rectangles at the right are similar. What side of rectangle *ABCD* corresponds to side *ZY*?**

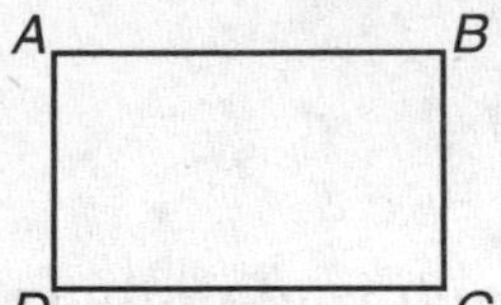

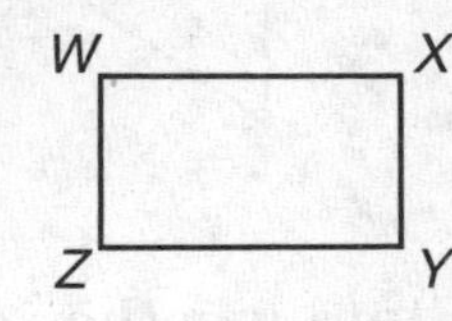

Corresponding sides represent the same side of similar figures. So, side *DC* corresponds to side *ZY*.

Exercises

Tell whether each pair of figures is *congruent*, *similar*, or *neither*.

1.

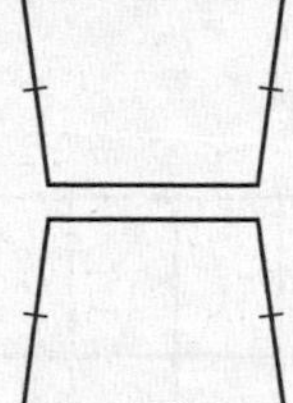

2.

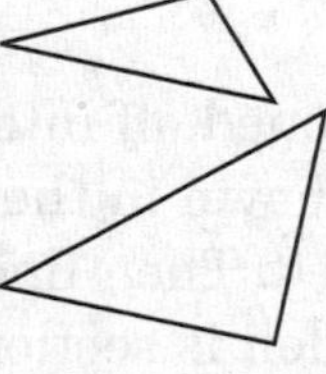

3.

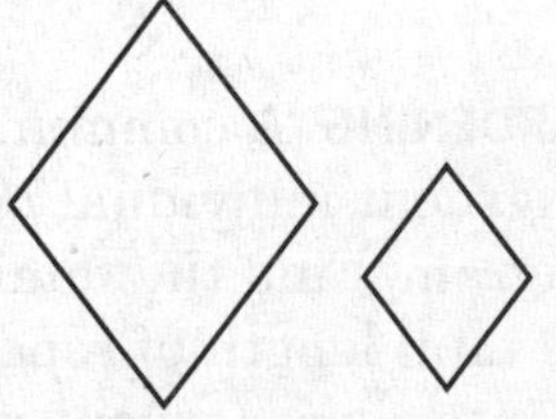

For Exercises 4 and 5, refer to the similar parallelograms at the right.

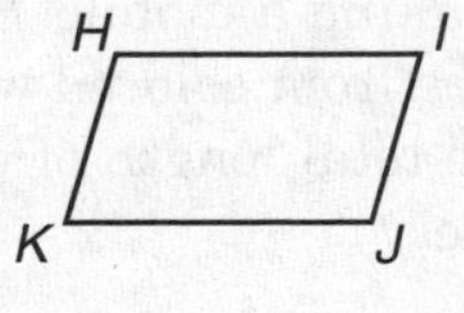

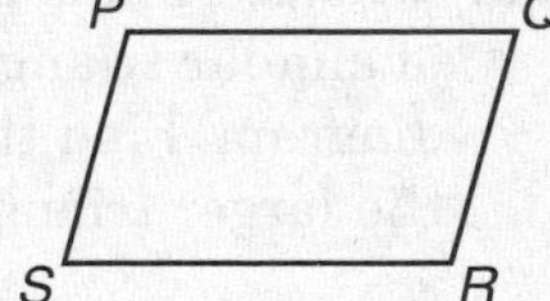

4. What side of parallelogram *HIJK* corresponds to side *QR*?

5. What side or parallelogram *PQRS* corresponds to side *HK*?

Lesson 9–7

Copyright © Glencoe/McGraw-Hill, a division of The McGraw-Hill Companies, Inc.

NAME ______________________ DATE __________ PERIOD ______

9-7 Practice

Similar and Congruent Figures

Tell whether each pair of figures is *congruent*, *similar*, or *neither*.

1.

2.

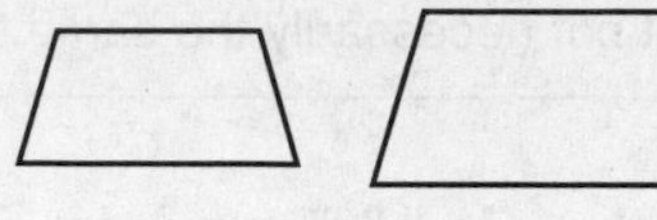

3.

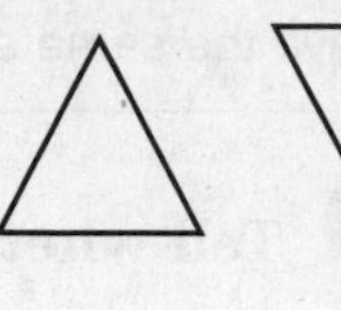

4.

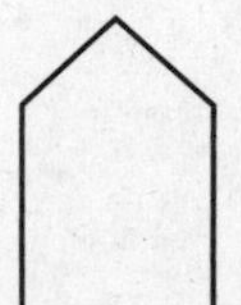

5.

6.

State whether each triangle is similar to triangle *RST*.

7.

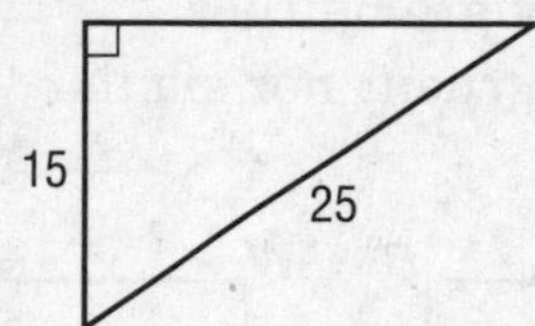

8.

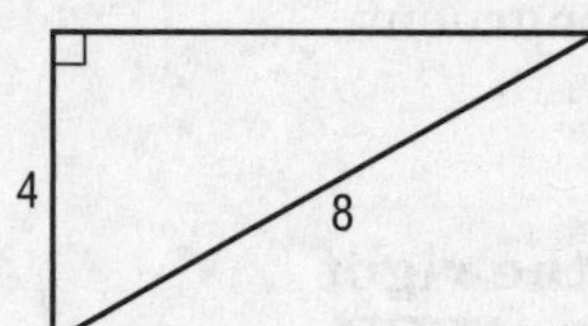

R S 6 10 T

State whether each rectangle is similar to rectangle *ABCD*.

9.

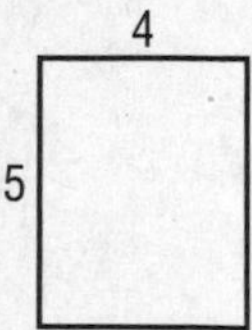

10.

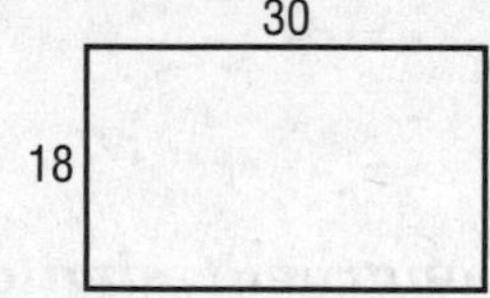

A 9 B 15 D C

11. GARDENING A community garden is sectioned off into 12 congruent individual plots with rope as shown in the diagram. Find the total length of rope used. Then find the total length of rope needed if the garden is sectioned off into six congruent rectangles.

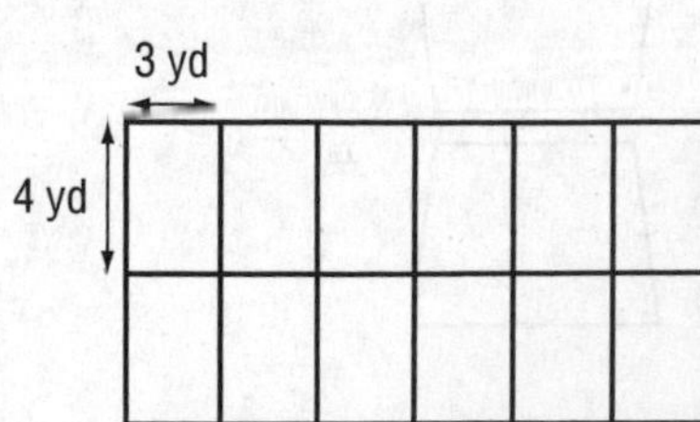

12. MOBILE Anh is making a mobile. She will make two sizes of similar triangles from colored wire as shown in the diagram. Find the total length of wire needed to make the larger triangle.

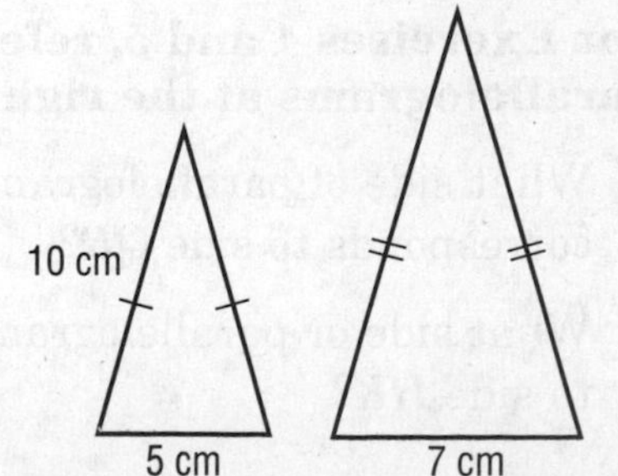

Copyright © Glencoe/McGraw-Hill, a division of The McGraw-Hill Companies, Inc.

NAME ______________________ DATE ____________ PERIOD ______

10-1 Study Guide and Intervention

Perimeter

The distance around any closed figure is called its **perimeter**. To find the perimeter, add the measures of all the sides of the figure.

Finding Perimeter		
Figure	**Words**	**Symbols**
Square	The perimeter *P* of a square is four times the measure of any of its sides *s*.	$P = 4s$
Rectangle	The perimeter *P* of a rectangle is the sum of the lengths and widths. It is also two times the length ℓ plus two times the width *w*.	$P = \ell + \ell + w + w$ $P = 2\ell + 2w$

Example 1 **Find the perimeter of the square.**

6 in.

$P = 4s$ Write the formula.
$P = 4(6)$ Replace *s* with 6.
$P = 24$ Multiply.

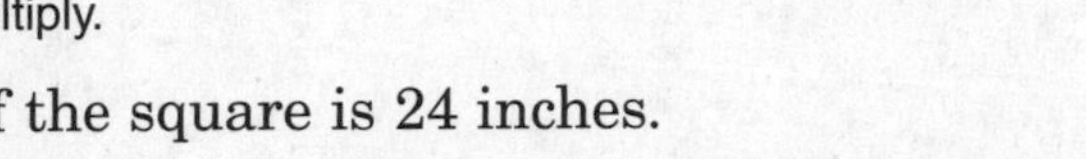

The perimeter of the square is 24 inches.

Example 2 **Find the perimeter of the rectangle.**

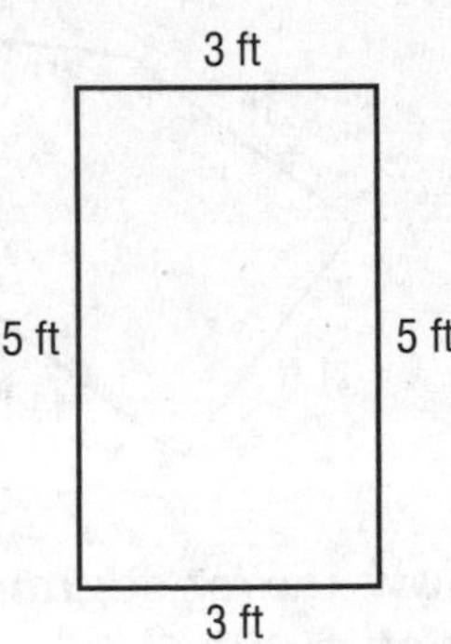

Estimate: $5 + 5 + 5 + 5 = 20$

$P = 2\ell + 2w$ Write the formula.
$P = 2(5) + 2(3)$ Replace ℓ with 5 and w with 3.
$P = 10 + 6$ Multiply.
$P = 16$ Add.

The perimeter of the rectangle is 16 feet. Compared to the estimate, the answer is reasonable.

Exercises

Find the perimeter of each square or rectangle.

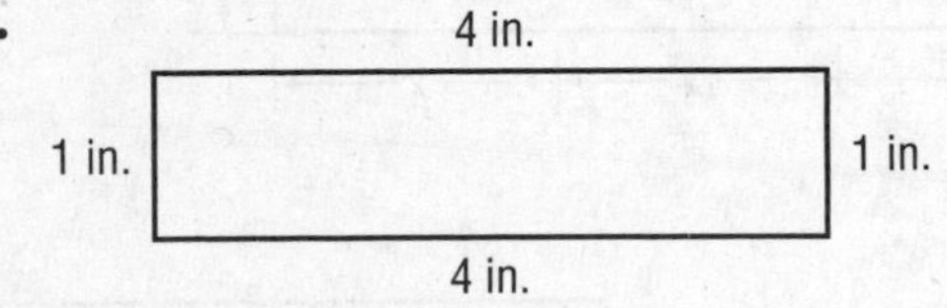

1. 4 in.; 1 in.; 1 in.; 4 in.

2.

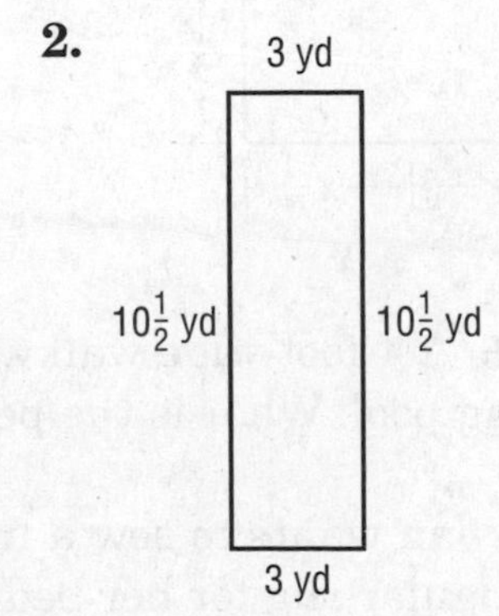

3. 5 ft; 5 ft

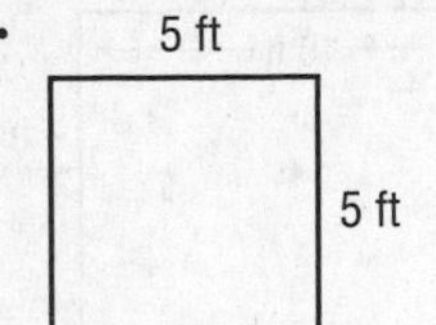

Copyright © Glencoe/McGraw-Hill, a division of The McGraw-Hill Companies, Inc.

NAME ______________________________ DATE ____________ PERIOD ______

10-1 Practice

Perimeter

Find the perimeter of each figure.

1.

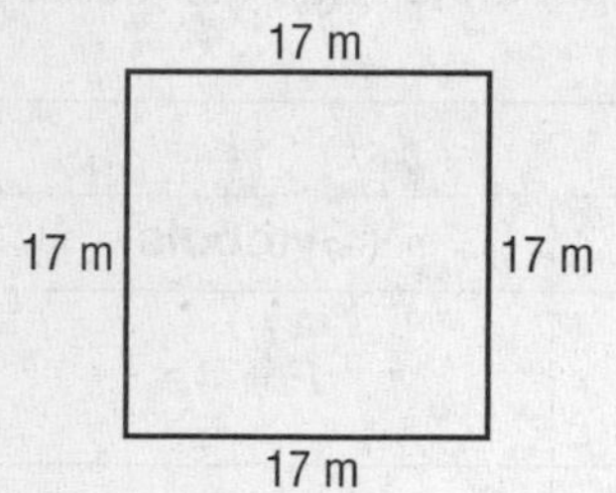

2.

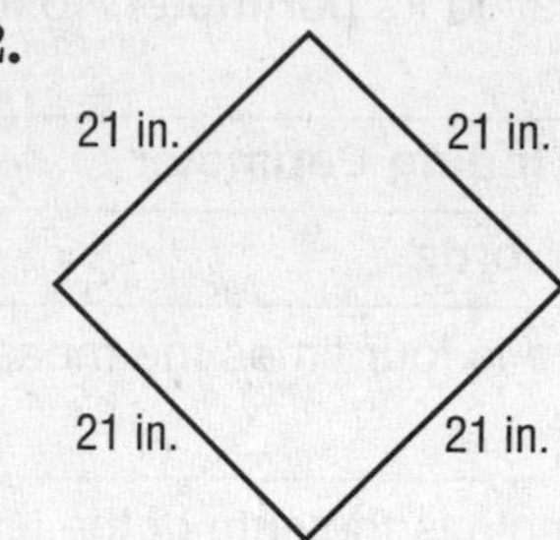

3.

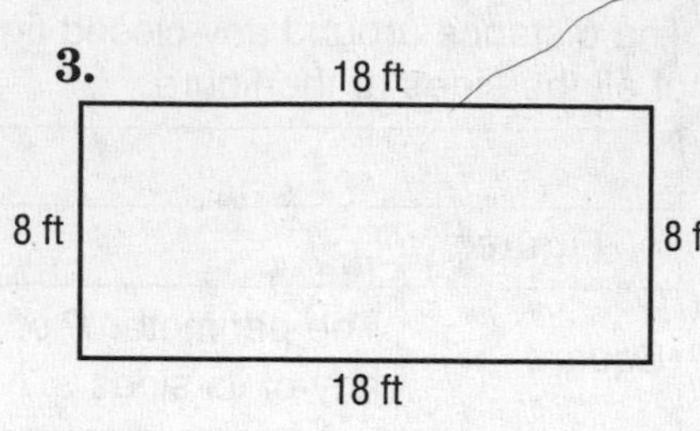

4.

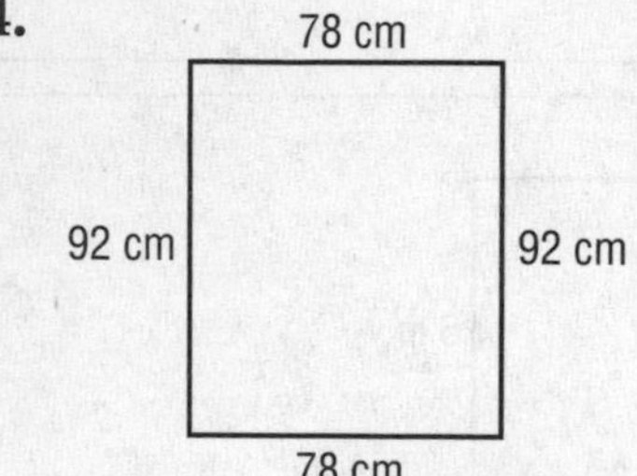

5.

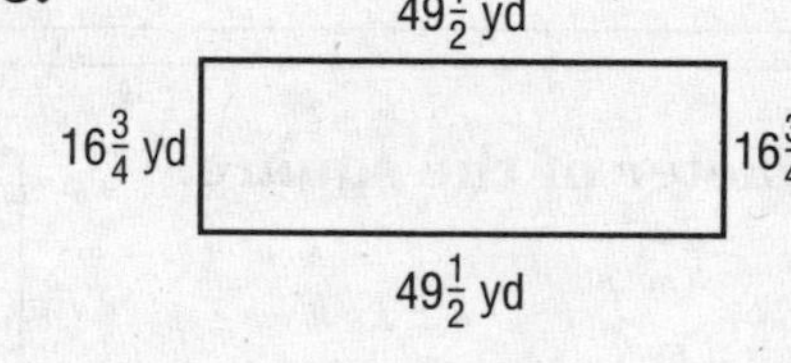

6.

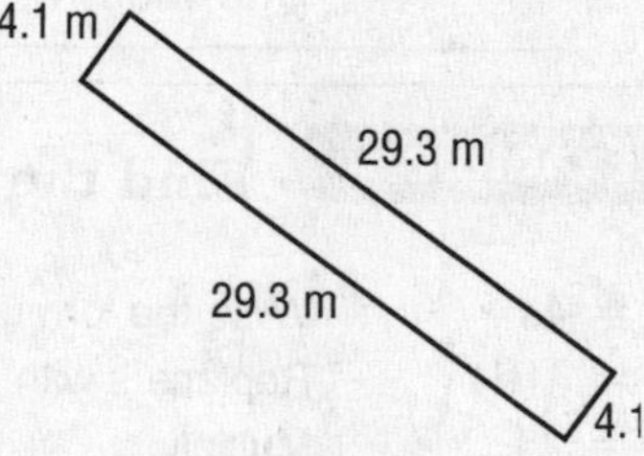

7.

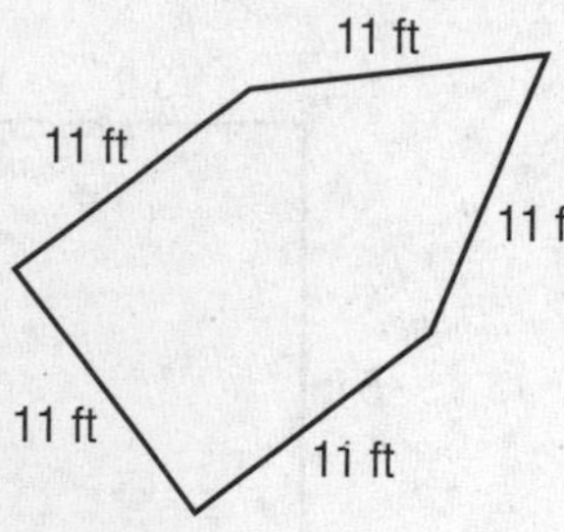

8.

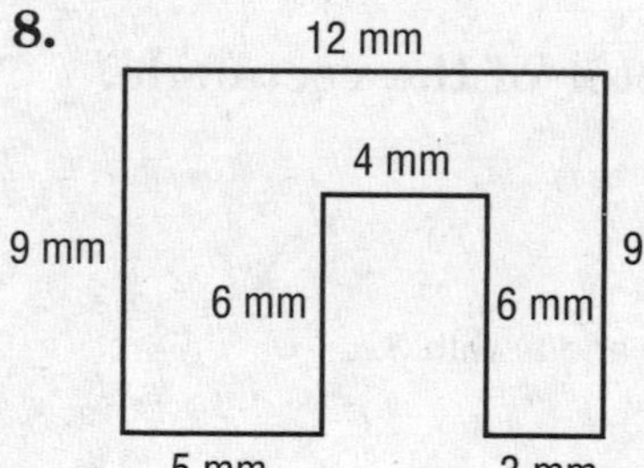

9.

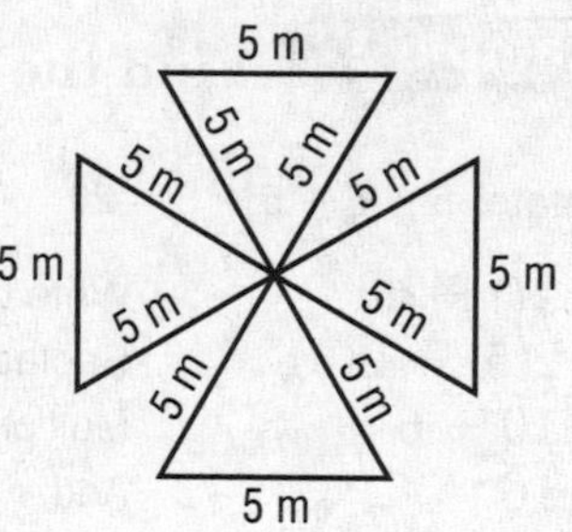

How many segments x units long are needed for the perimeter of each figure?

10.

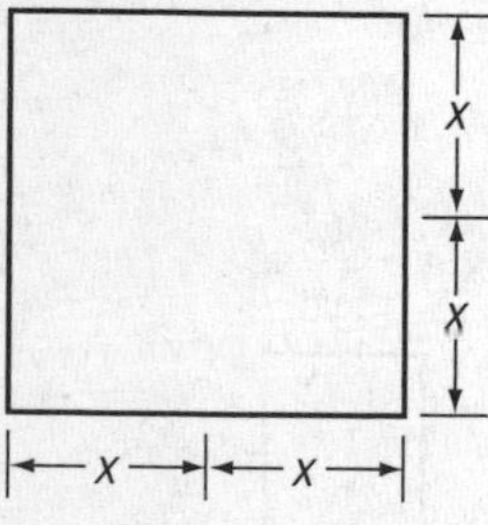

11.

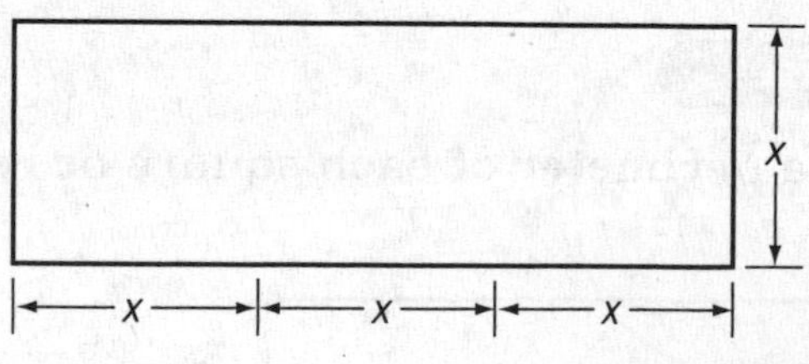

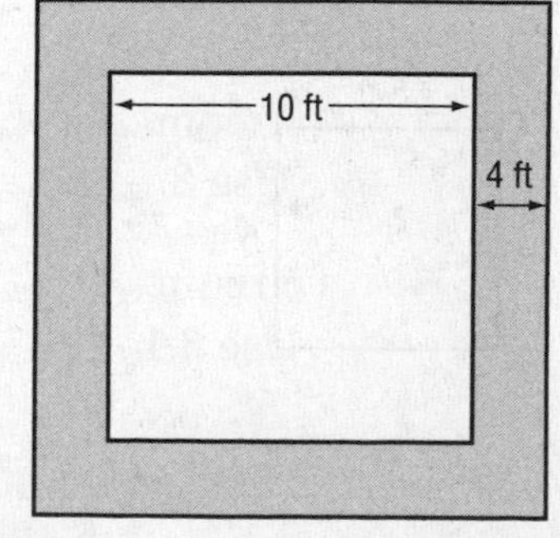

12. POOLS A 4-foot wide walkway surrounds a 10-foot square wading pool. What is the perimeter of the walkway?

13. RUGS Jan wants to sew a fringe border on all sides of a rectangular rug for her bedroom. The rug is 3.4 feet wide and 5.5 feet long. How many feet of fringe does she need?

Copyright © Glencoe/McGraw-Hill, a division of The McGraw-Hill Companies, Inc.

NAME ______________________ DATE __________ PERIOD _____

10-2 Study Guide and Intervention

Circles and Circumference

Lesson 10–2

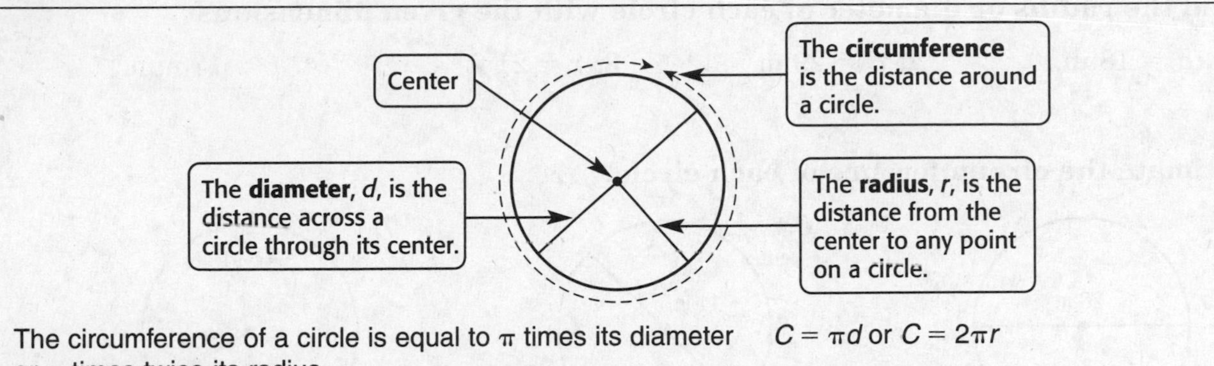

The circumference of a circle is equal to π times its diameter or π times twice its radius. $C = \pi d$ or $C = 2\pi r$

Example 1 **Estimate the circumference of a circle whose diameter is 4 meters.**

$C = \pi d$ Write the formula.
$\approx 3 \times 4$ Replace π with 3 and d with 4.
≈ 12 Multiply.

The circumference of the circle is about 12 meters.

Example 2 **Find the circumference of a circle whose radius is 13 inches. Use 3.14 for π. Round to the nearest tenth.**

$C = 2\pi r$ Write the formula.
$= 2 \times 3.14 \times 13$ Replace r with 13 and π with 3.14.
$= 81.64$ Multiply.

Rounded to the nearest tenth, the circumference is about 81.6 inches.

Exercises

Estimate the circumference of each circle.

1.

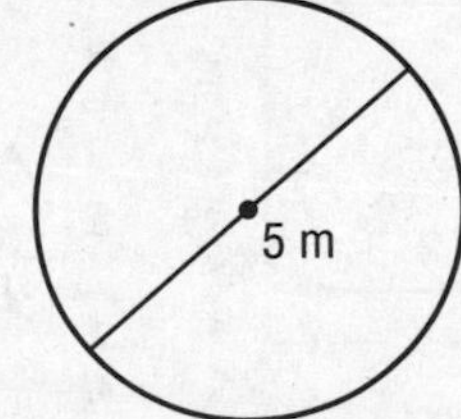

2.

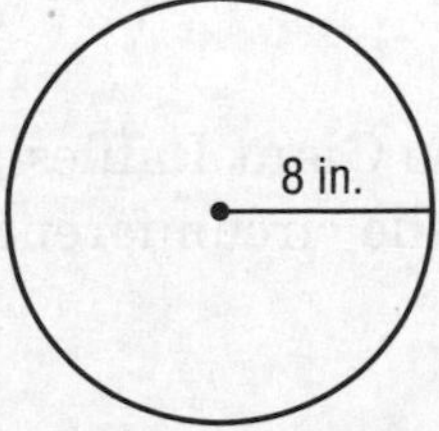

3.

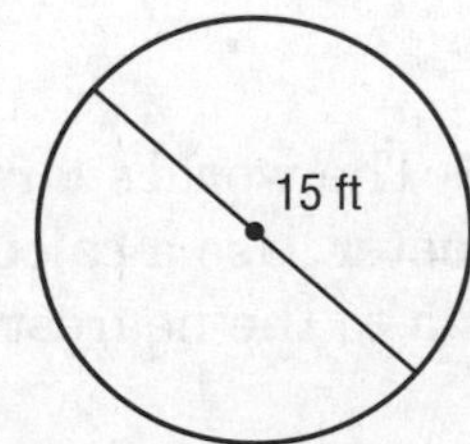

4. The radius of a circle measures 16 miles. Find the measure of its circumference to the nearest tenth. Use 3.14 for π.

5. Find the circumference of a circle whose diameter is 12 yards. Use 3.14 for π. Round to the nearest tenth.

6. Find the circumference of a circle with a radius of 7 inches. Use 3.14 for π. Round to the nearest tenth.

Copyright © Glencoe/McGraw-Hill, a division of The McGraw-Hill Companies, Inc.

NAME ______________________ DATE __________ PERIOD ______

10-2 Practice

Circles and Circumference

Find the radius or diameter of each circle with the given dimensions.

1. $d = 18$ in. **2.** $d = 29$ m **3.** $r = 21$ ft **4.** $r = 13$ mm

Estimate the circumference of each circle.

5.

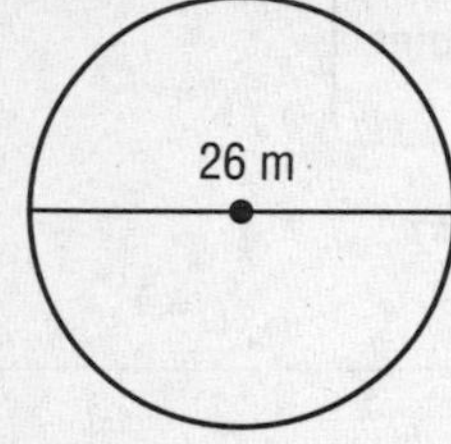

6.

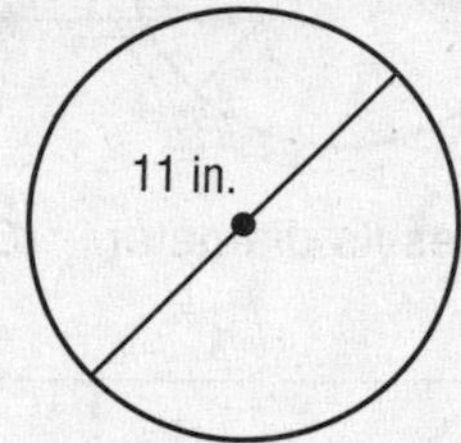

7.

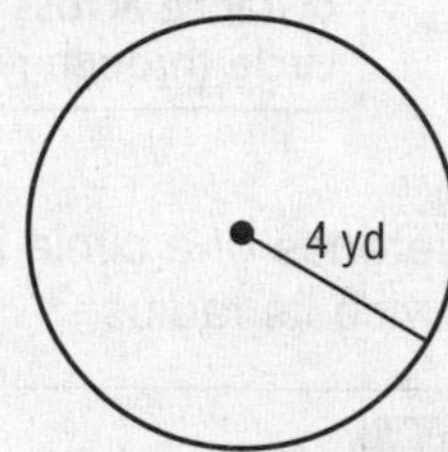

8. $d = 31$ mm **9.** $r = 29$ cm **10.** $d = 32$ yd

Find the circumference of each circle. Use 3.14 for π. Round to the nearest tenth.

11.

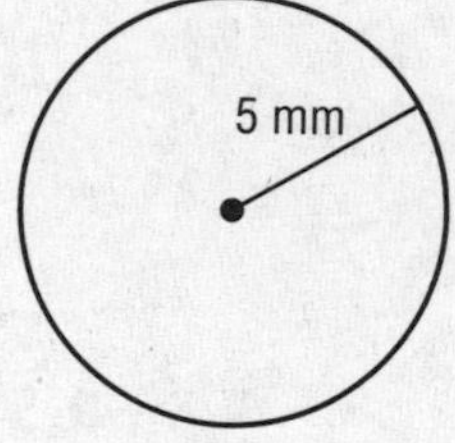

12.

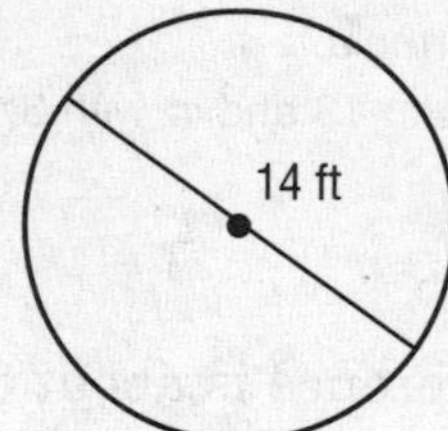

13.

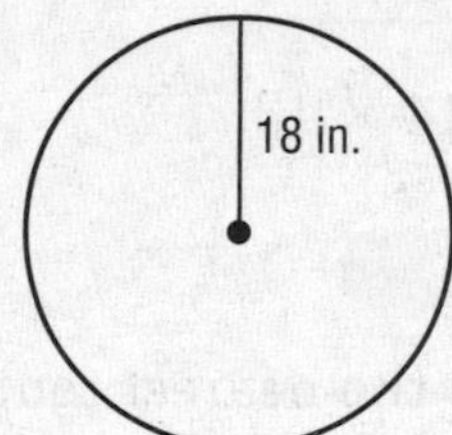

14. $r = 22$ cm **15.** $r = 15$ yd **16.** $d = 31$ m

17. PLANTS The world's largest flower, the Giant Rafflesia, is 91 centimeters in diameter. Use a calculator to find the circumference of a Giant Rafflesia to the nearest tenth.

18. GEOLOGY Ubehebe Crater in Death Valley has a diameter of a little more than $\frac{1}{2}$ mile. If Latisha walks around its rim at a rate of 2 miles per hour, about how long will it take her to walk all the way around the crater? Find your answer to the nearest tenth. Use 3.14 for π.

Copyright © Glencoe/McGraw-Hill, a division of The McGraw-Hill Companies, Inc.

NAME ______________________ DATE ____________ PERIOD ____

10-3 Study Guide and Intervention

Area of Parallelograms

The area A of a parallelogram is the product of any base b and its height h.

Symbols $A = bh$

Model

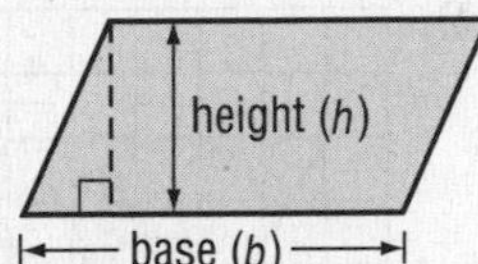

Examples **Find the area of each parallelogram.**

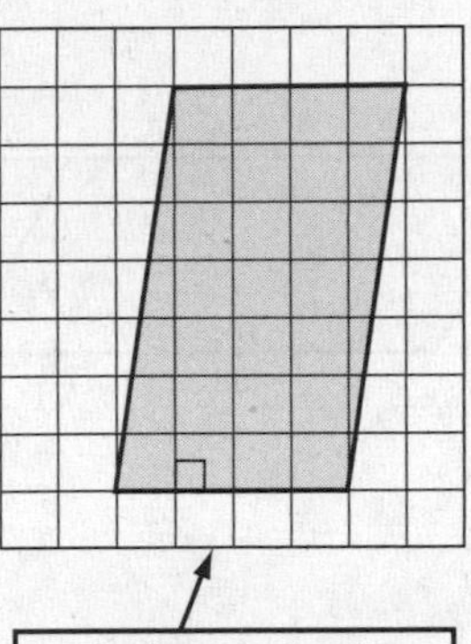

The base is 4 units, and the height is 7 units.

$A = bh$

$A = 4 \times 7$

$A = 28$

The area is 28 square units or 28 units^2.

2

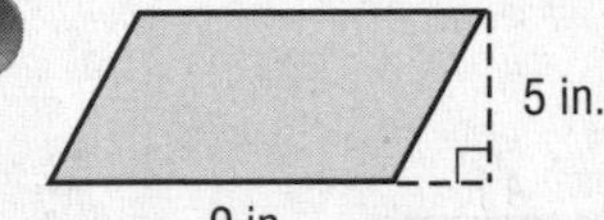

$A = bh$

$A = 9 \times 5$

$A = 45$

The area is 45 square inches or 45 in^2.

Exercises

Find the area of each parallelogram.

1.

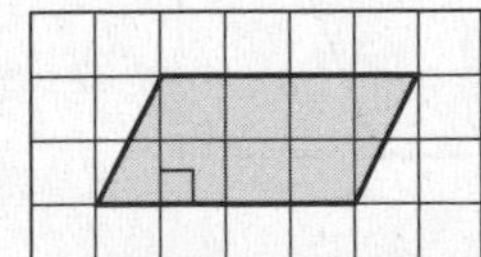

2.

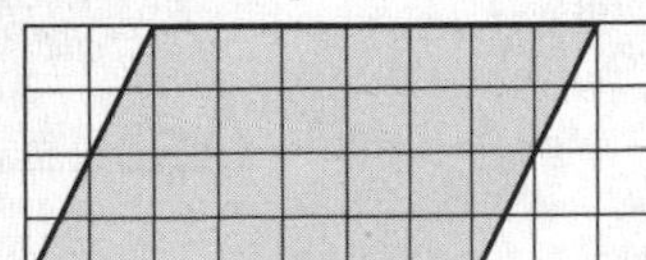

3.

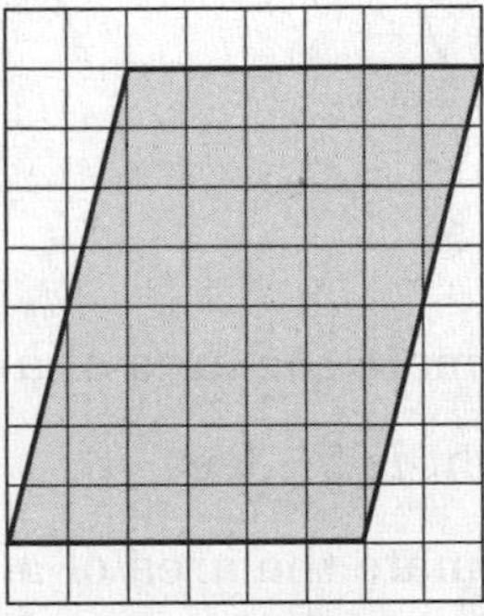

4.

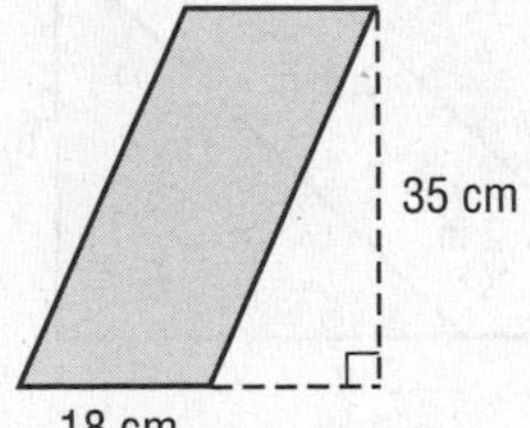

5.

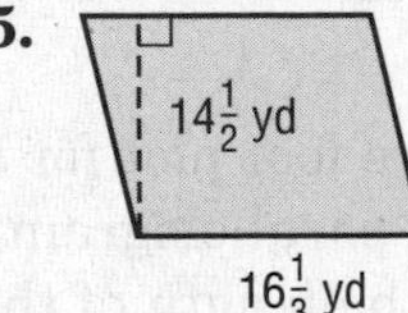

6.

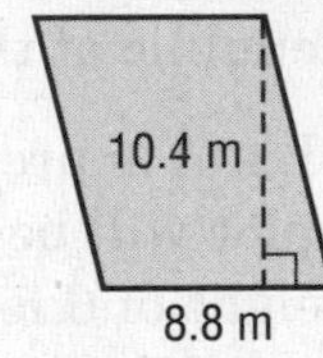

Lesson 10–3

Copyright © Glencoe/McGraw-Hill, a division of The McGraw-Hill Companies, Inc.

NAME ______________________________ DATE ____________ PERIOD _____

10-3 Practice

Area of Parallelograms

Find the area of each parallelogram.

1.

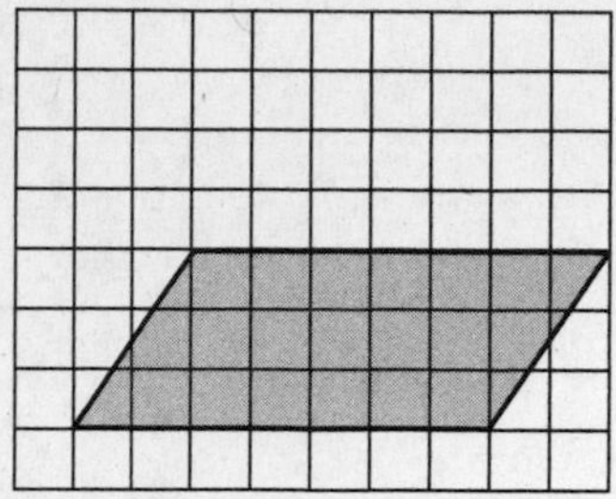

2.

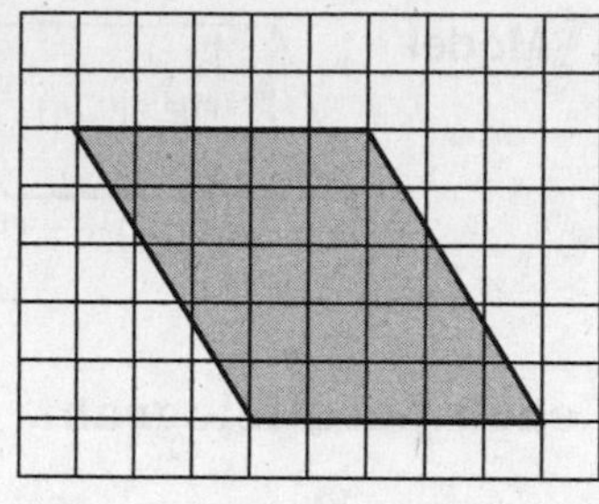

3.

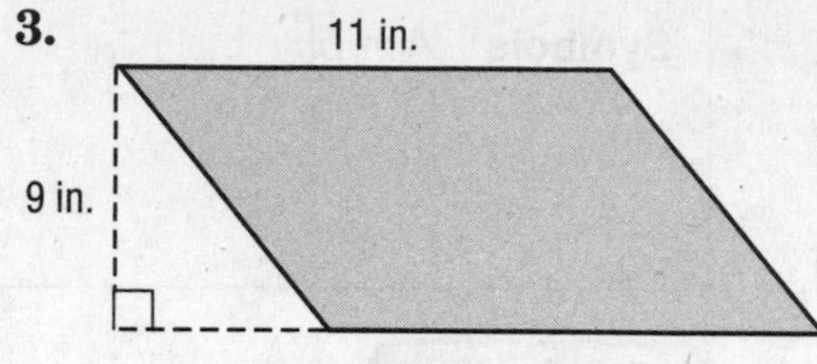

4.

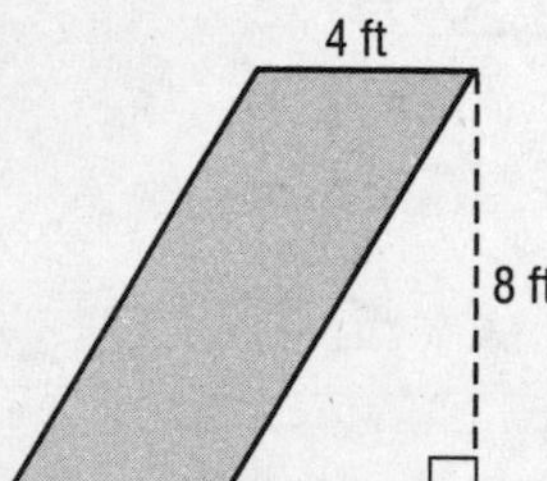

5.

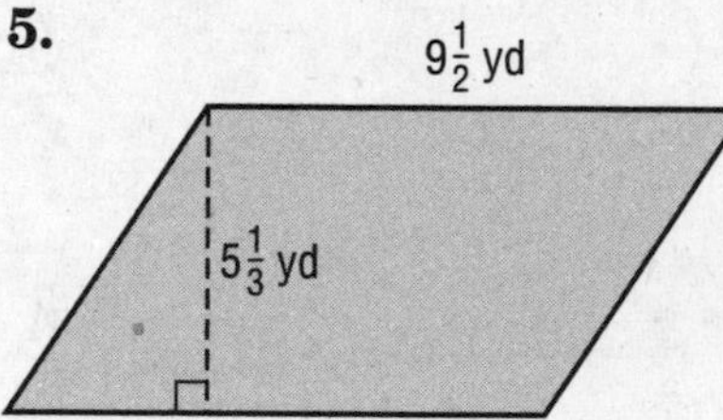

6.

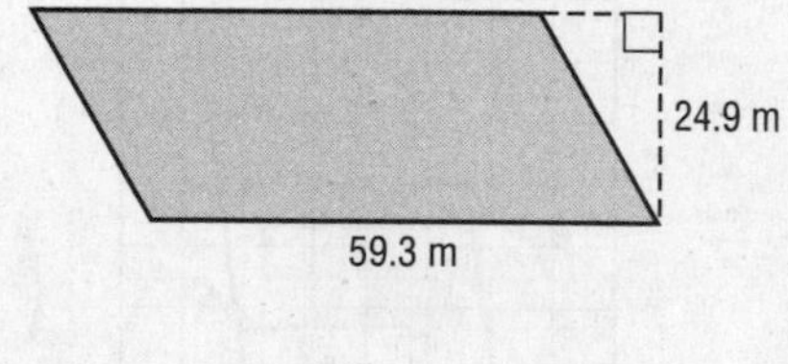

Find the area of the shaded region in each figure.

7.

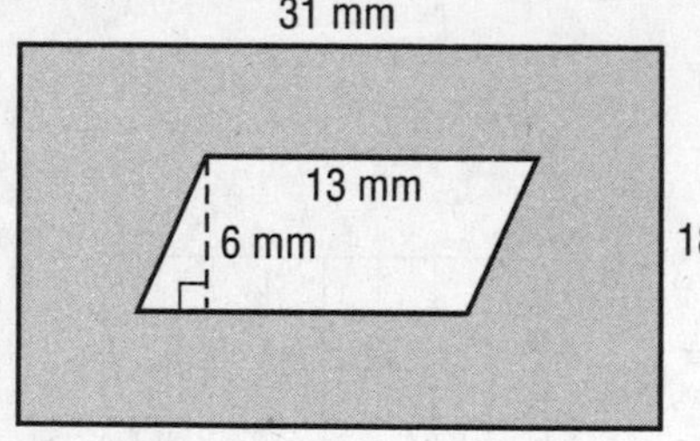

8.

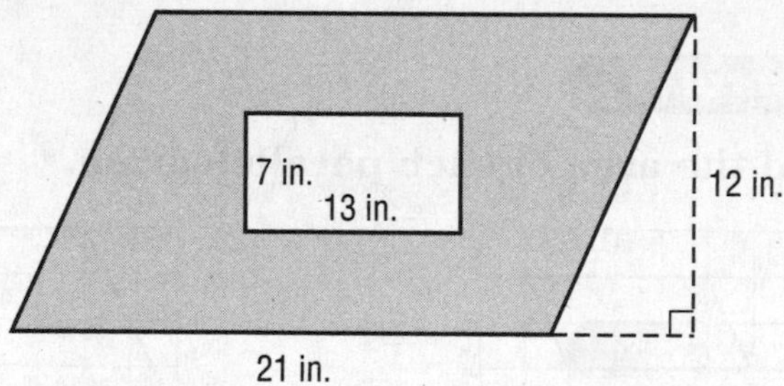

9. Estimate the area of a parallelogram whose base is $6\frac{5}{8}$ feet and whose height is $5\frac{2}{5}$ feet.

10. Estimate the area of a parallelogram with base 9.44 yards and height 7.56 yards.

11. **FLAGS** Estimate the area of the shaded region of the flag of the Republic of the Congo.

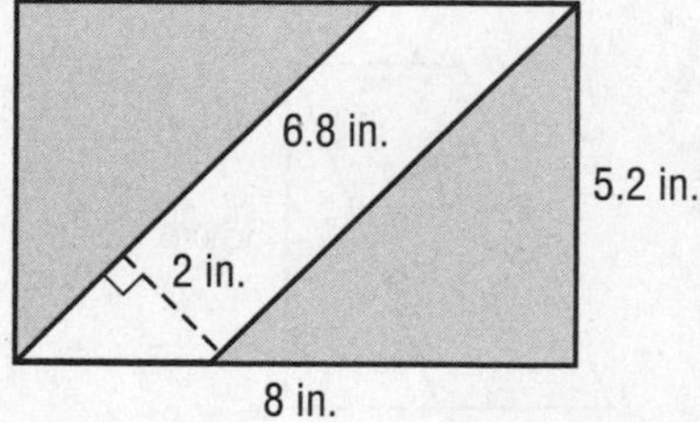

12. **GARDENING** Liam is preparing a 78 square foot plot for a garden. The plot will be in the shape of a parallelogram that has a height of 6 feet. What will be the length of the base of the parallelogram? Explain your reasoning.

Copyright © Glencoe/McGraw-Hill, a division of The McGraw-Hill Companies, Inc.

NAME ______________________ DATE ____________ PERIOD _____

10-4 Study Guide and Intervention

Area of Triangles

The area *A* of a triangle is one half the product of any base *b* and its height *h*.

Symbols $A = \frac{bh}{2}$ **Model**

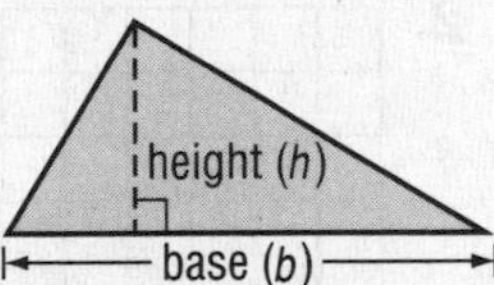

Examples **Find the area of each triangle.**

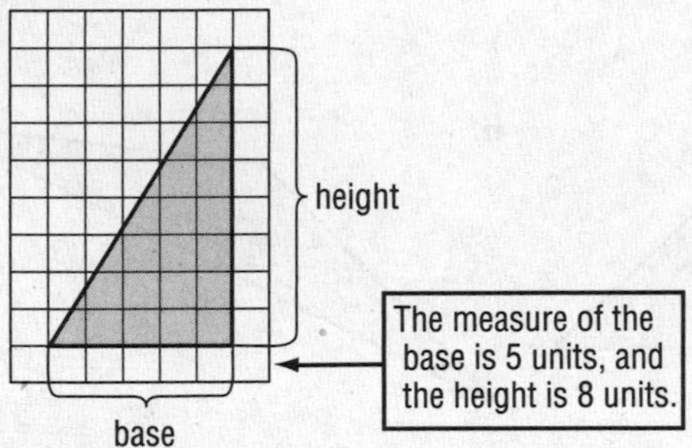

$A = \frac{bh}{2}$ Area of a triangle

$A = \frac{5 \times 8}{2}$ Replace *b* with 5 and *h* with 8.

$A = \frac{40}{2}$ Simplify the numerator.

$A = 20$ Divide.

The area of the triangle is 20 square units.

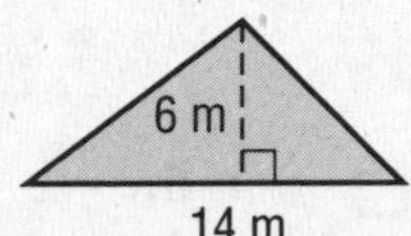

$A = \frac{bh}{2}$ Area of a triangle

$A = \frac{14 \times 6}{2}$ Replace *b* with 14 and *h* with 6.

$A = \frac{84}{2}$ Simplify the numerator.

$A = 42$ Divide.

The area of the triangle is 42 square meters.

Exercises

Find the area of each triangle.

1.

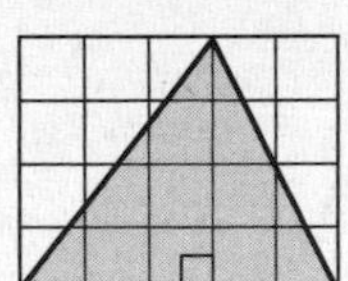

2.

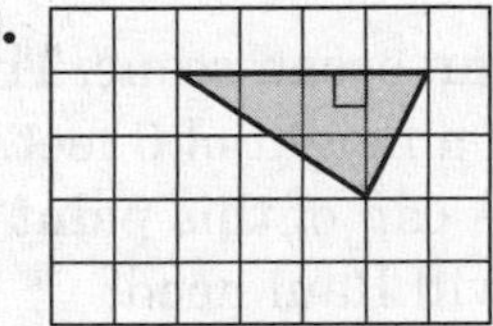

3.

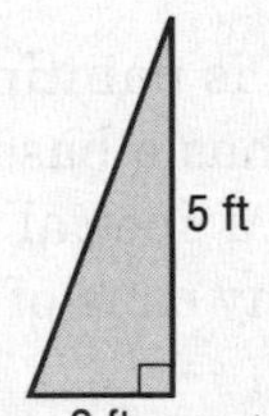

4.

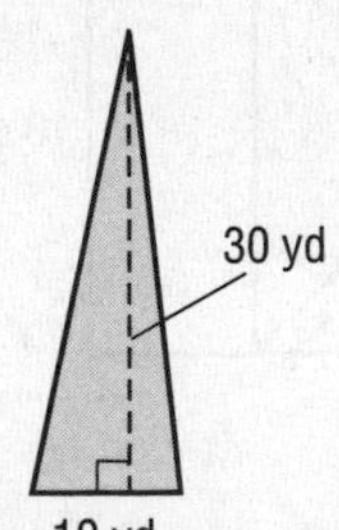

5.

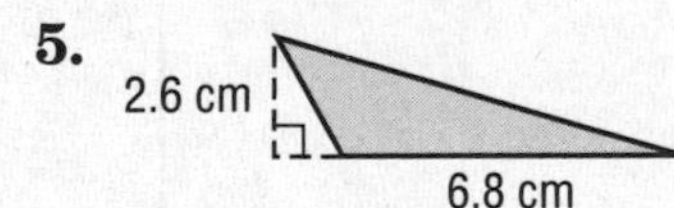

6.

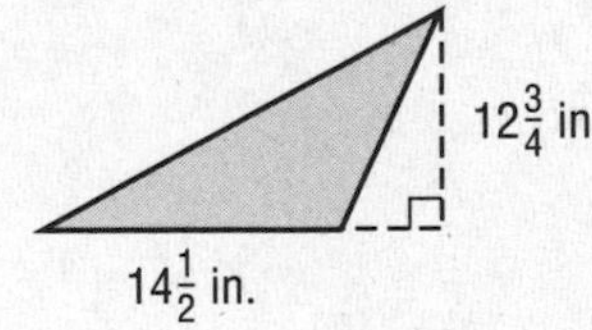

Copyright © Glencoe/McGraw-Hill, a division of The McGraw-Hill Companies, Inc.

Lesson 10–4

NAME ______________________ DATE __________ PERIOD ______

10-4 Practice

Area of Triangles

Find the area of each triangle.

1.

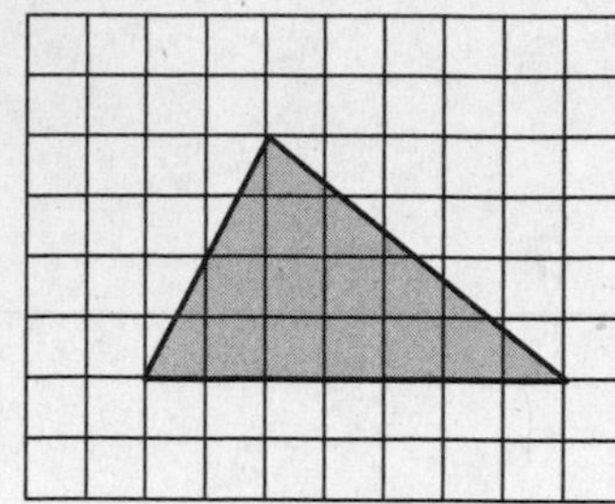

2.

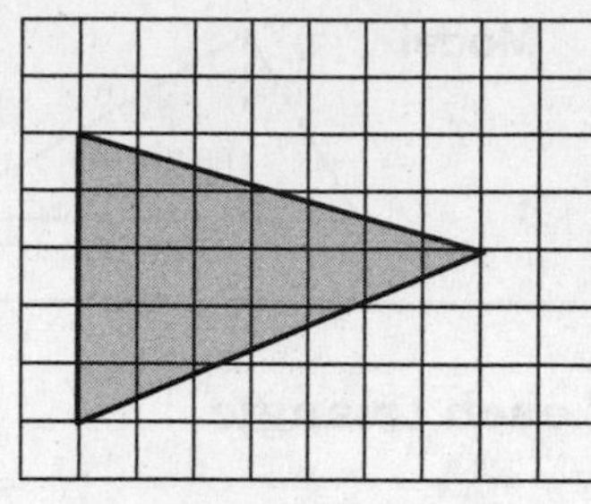

3.

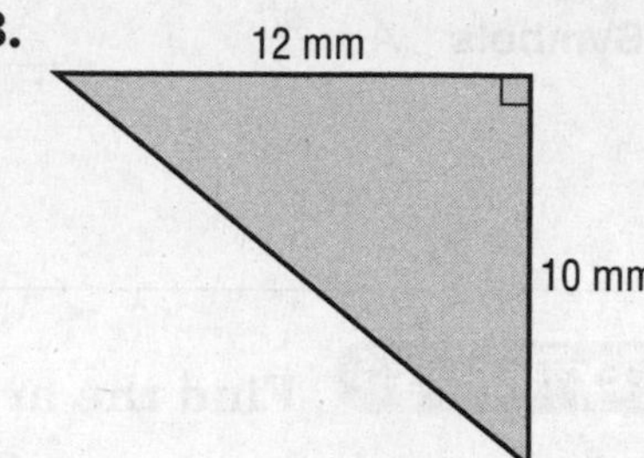

4.

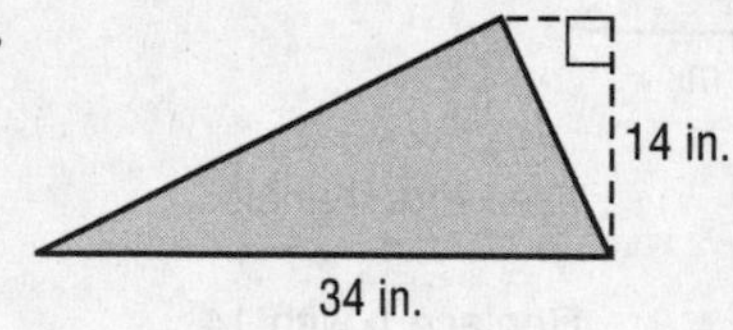

5.

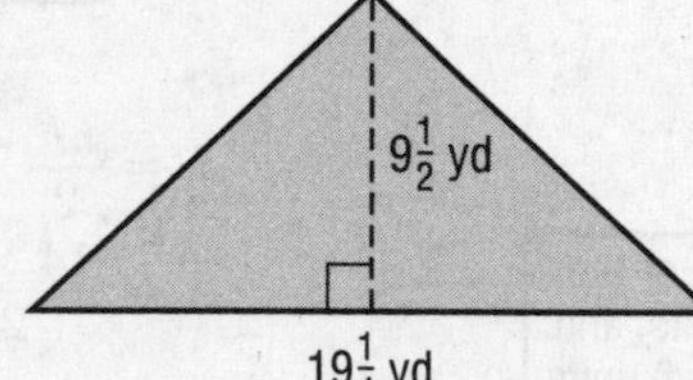

6.

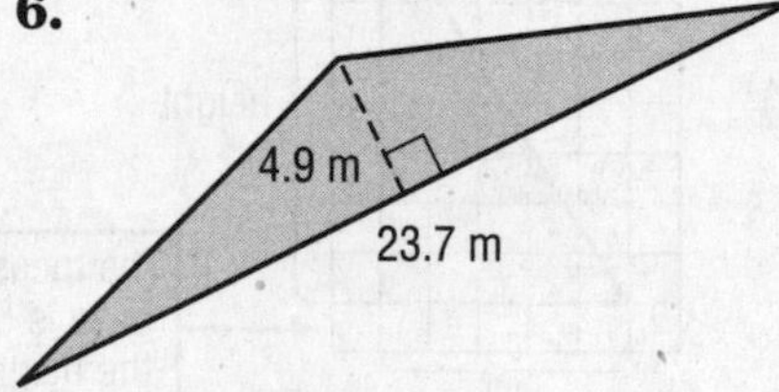

7. height: 15 ft
base: 38 ft

8. height: 22 cm
base: 17 cm

9. height: 12 in.
base: 21 in.

10. COMPLEX FIGURES Find the area of the figure at the right.

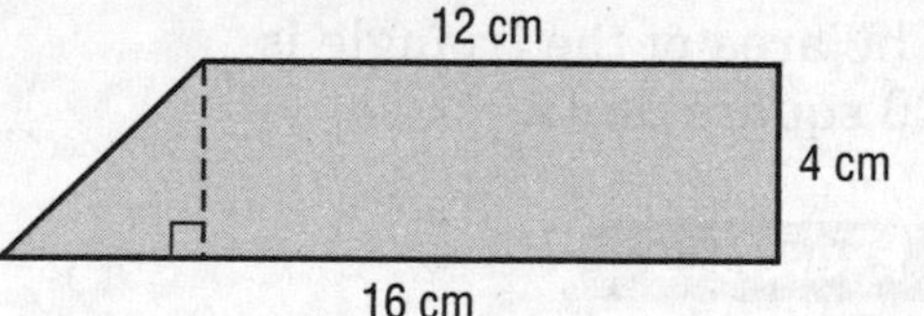

11. MURALS Raul is painting a mural of an ocean scene. The triangular sail on a sailboat has a base of 4 feet and a height of 6 feet. Raul will paint the sail using a special white paint. A can of this paint covers 10 square feet. How many cans of white paint will Raul need?

12. FLAGS What is the area of the triangle on the flag of Bosnia and Herzegovina?

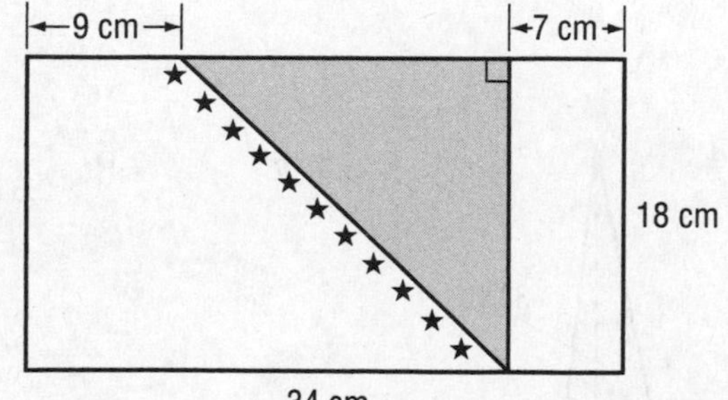

Copyright © Glencoe/McGraw-Hill, a division of The McGraw-Hill Companies, Inc.

NAME ______________________________ DATE ____________ PERIOD _____

10-5 Study Guide and Intervention

Problem-Solving Investigation: Make a Model

When solving problems, one strategy that is helpful is to *make a model.* If a problem gives data that can be displayed visually, it may be useful to make a model of the situation. The model can then be used in order to solve the problem.

You can use the *make a model* strategy, along with the following four-step problem solving plan to solve a problem.

1 Understand – Read and get a general understanding of the problem.

2 Plan – Make a plan to solve the problem and estimate the solution.

3 Solve – Use your plan to solve the problem.

4 Check – Check the reasonableness of your solution.

Example DISPLAYS **A grocery store employee is making a pyramid display of boxes of a new cereal. If he doesn't want to have more than 4 rows in his display, what is the least number of cereal boxes he can use?**

Understand The cereal boxes need to be stacked in the shape of a pyramid. There should only be 4 rows in the pyramid. We need to know the minimum number of boxes of cereal needed to make a pyramid.

Plan Make a model to find the number of cereal boxes needed.

Solve Use a rectangle to represent each cereal box.

The least number of boxes needed is 4 + 3 + 2 + 1, or 10 boxes.

Check Count the number of boxes in the model. There are 10 boxes.

Exercise

TILING Michael has 18 decorative square tiles to make a design on a kitchen backsplash. He wants to arrange them in a rectangular shape with the least perimeter possible. How many tiles will be in each row?

Lesson 10–5

Copyright © Glencoe/McGraw-Hill, a division of The McGraw-Hill Companies, Inc.

10-5 Practice

Problem-Solving Investigation: Make a Model

Mixed Problem Solving

Use the make a model strategy to solve Exercises 1 and 2.

1. **QUILTING** Ms. Mosely is sewing together blocks of fabric in a pattern of small squares and triangles to make a quilt that is 3 feet square. How many small squares will she need? How many small triangles will she need?

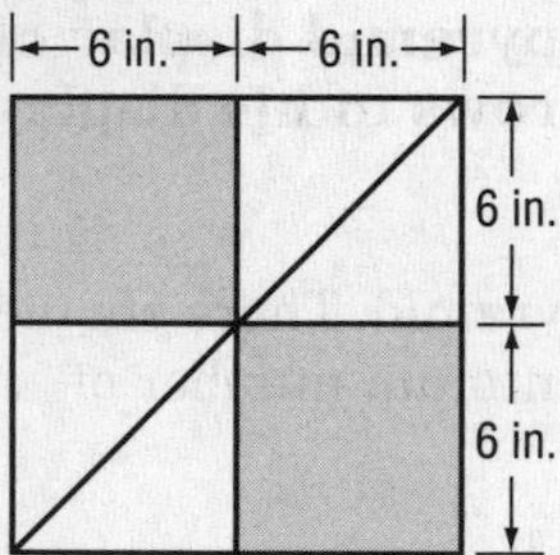

2. **DISPLAY** Anaba is stacking cereal boxes in a pyramid-shaped display. The bottom layer has 10 boxes. There are two fewer boxes in each layer than the layer below. How many boxes are in the display?

Use any strategy to solve Exercises 3–6. Some strategies are shown below.

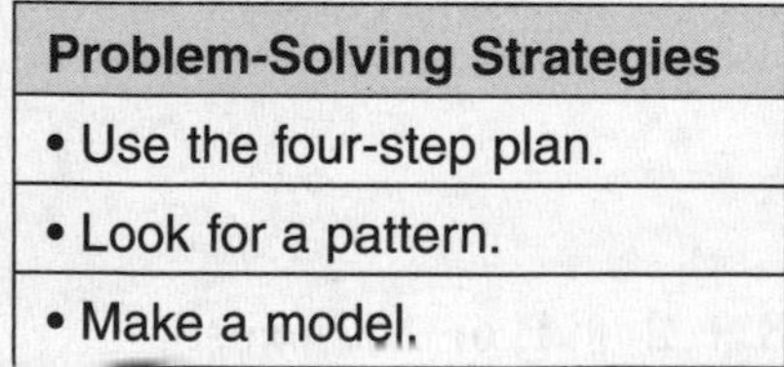

Problem-Solving Strategies
• Use the four-step plan.
• Look for a pattern.
• Make a model.

3. **PATTERNS** Draw the next figure.

4. **ART** Kris folded a piece of construction paper into thirds and then in half. He punched a hole through all layers. How many holes will there be when he unfolds the paper?

5. **LOANS** Mr. Kartini bought a boat on credit. His loan, including interest, is $9,860. If he makes monthly payments of $85, how many years will it take him to pay off the loan?

6. **MUSIC** Refer to the graph. How many fewer girls took band class in 2005 than in 2004?

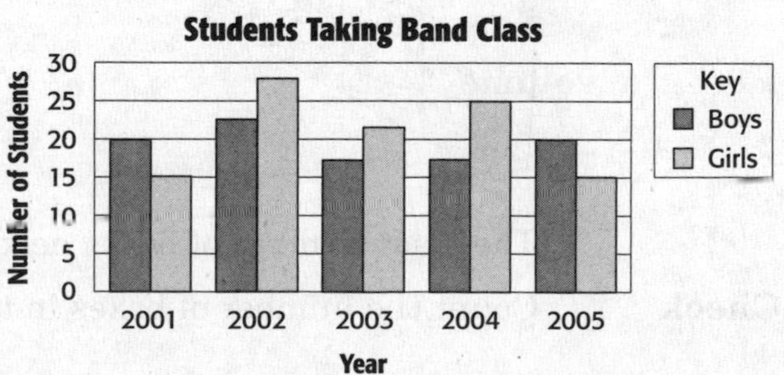

Copyright © Glencoe/McGraw-Hill, a division of The McGraw-Hill Companies, Inc.

NAME ______________________ DATE __________ PERIOD ____

10-6 Study Guide and Intervention

Volume of Rectangular Prisms

The amount of space inside a three-dimensional figure is the **volume** of the figure. Volume is measured in **cubic units**. This tells you the number of cubes of a given size it will take to fill the prism.

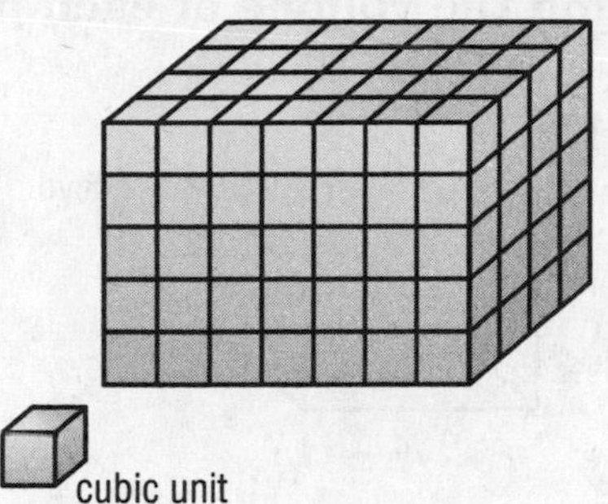

The volume V of a rectangular prism is the product of its length ℓ, width w, and height h.
Symbols $V = \ell wh$

Model

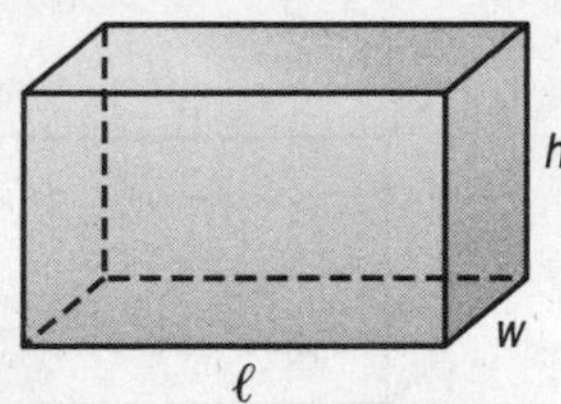

You can also multiply the area of the base B by the height h to find the volume V.
Symbols $V = Bh$

Model

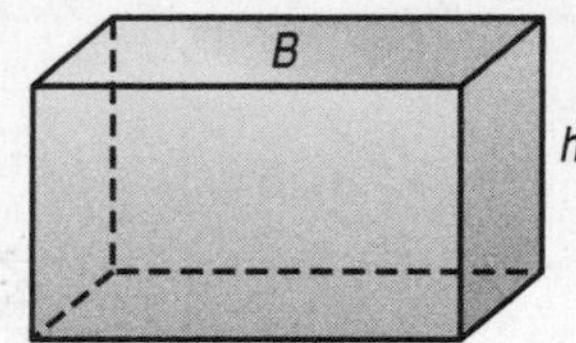

Example **Find the volume of the rectangular prism.**

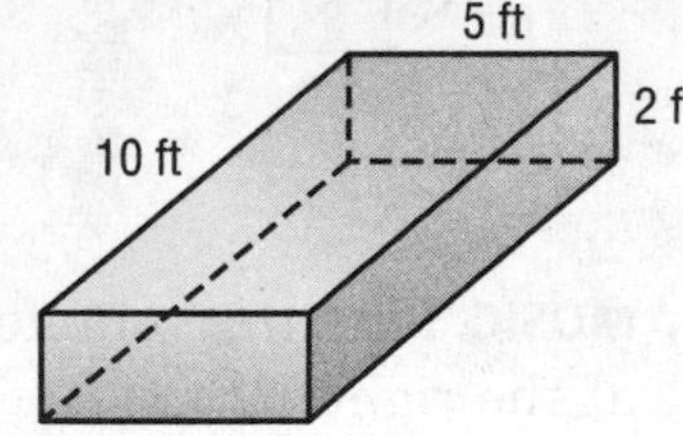

Method 1 Use $V = \ell wh$.

$V = \ell wh$

$V = 10 \times 5 \times 2$

$V = 100$

The volume is 100 ft^3.

Method 2 Use $V = Bh$.

$V = Bh$ — B, the area of the base, is 10×5 or 50.

$V = 50 \times 2$

$V = 100$

The volume is 100 ft^3.

Exercises

Find the volume of each prism.

1.

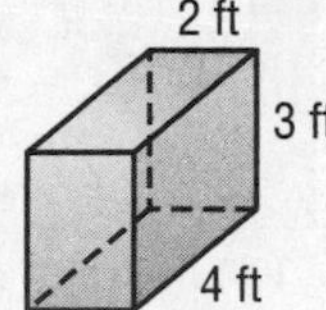

2.

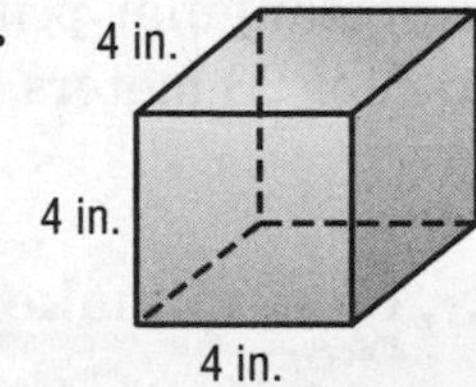

3.

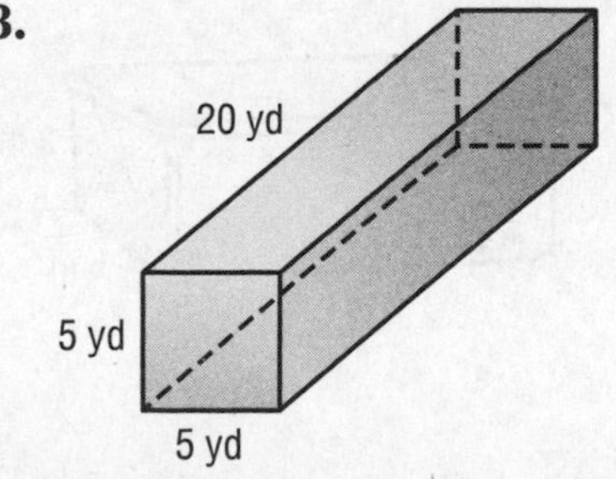

4.

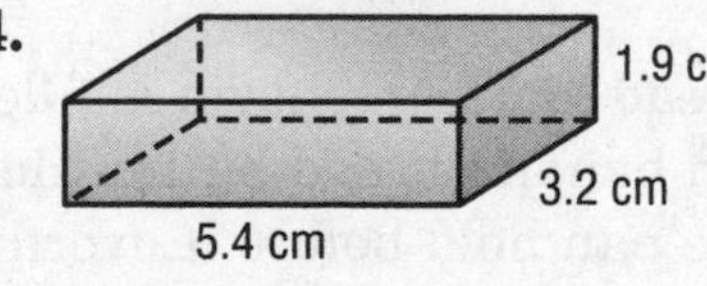

Lesson 10–6

Copyright © Glencoe/McGraw-Hill, a division of The McGraw-Hill Companies, Inc.

NAME ______________________ DATE ____________ PERIOD _____

10-6 Practice

Volume of Rectangular Prisms

Find the volume of each prism.

1.

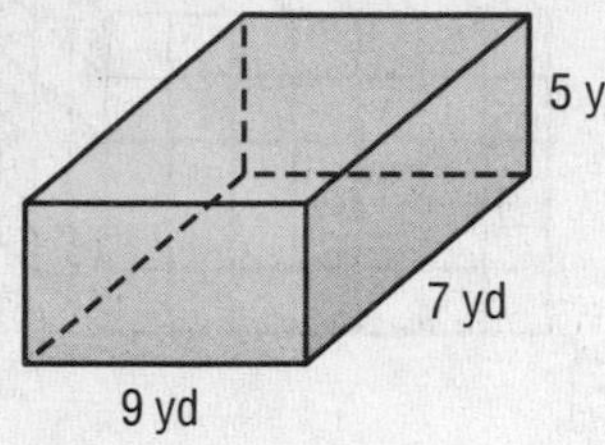

2.

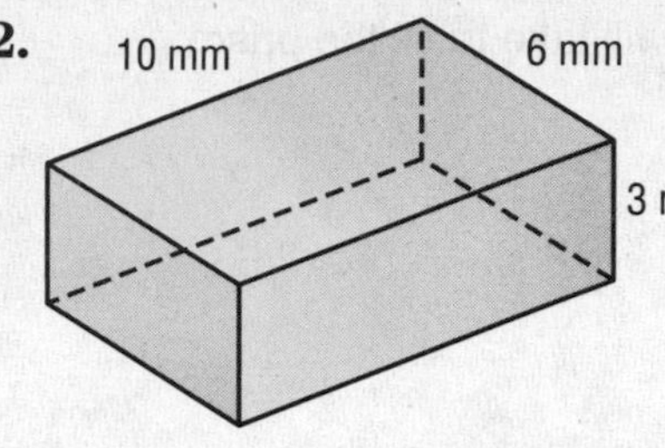

3.

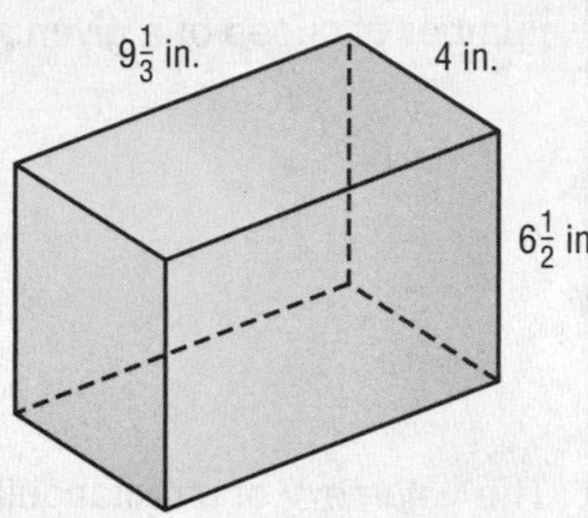

4.

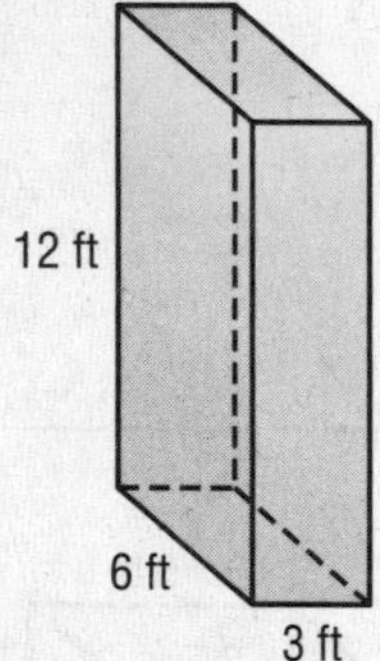

5.

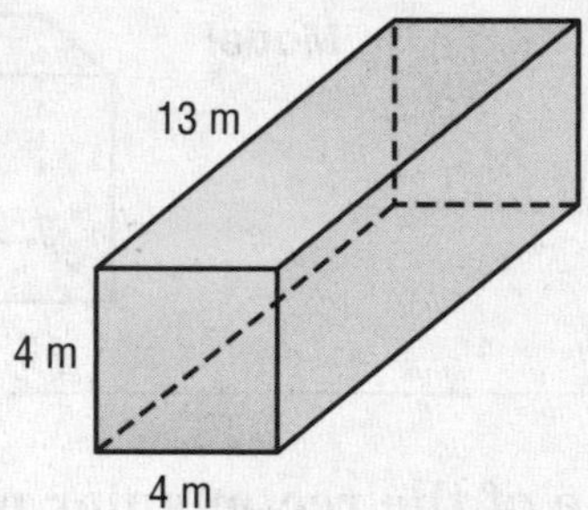

6.

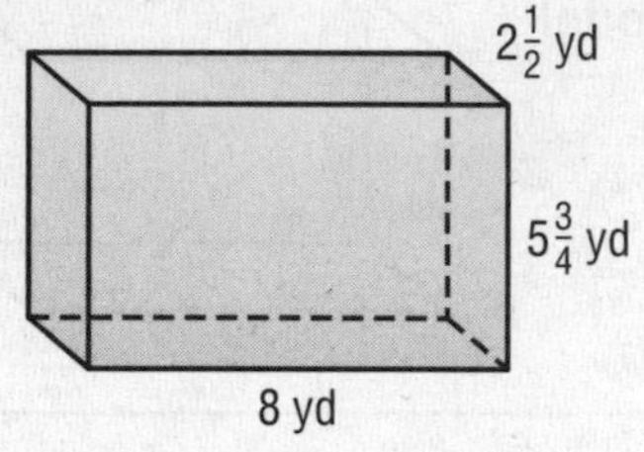

7. MUSIC Find the volume of the CD box shown at the right.

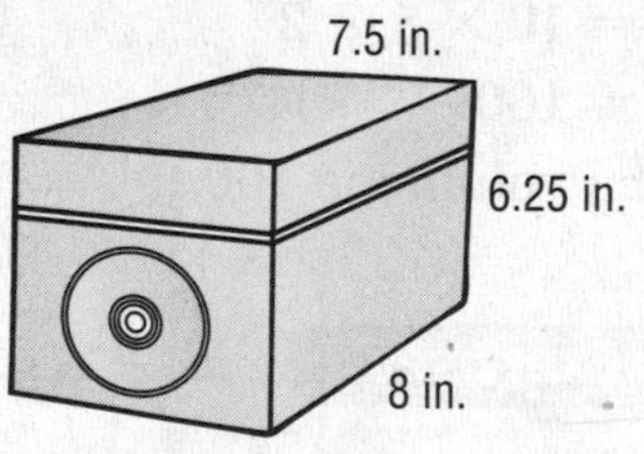

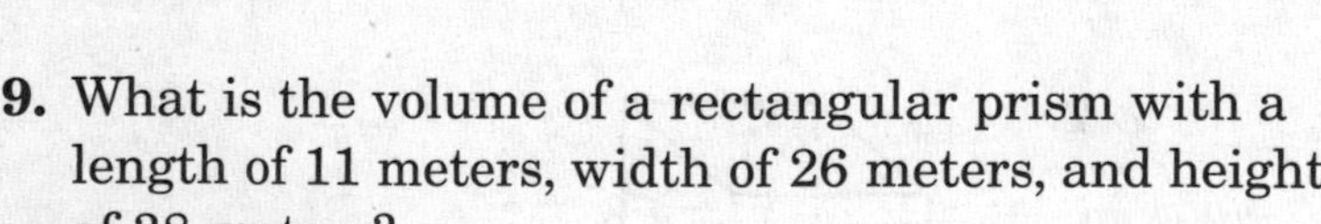

8. TOYS Jamie's younger brother has a toy box that is 3.6 feet long, 2.4 feet wide, and 4.5 feet high. What is the volume of toy box?

9. What is the volume of a rectangular prism with a length of 11 meters, width of 26 meters, and height of 38 meters?

Replace each ● with <, >, or = to make a true sentence.

10. 1 yd^3 ● 3 ft^3 **11.** 1 yd^3 ● 27 ft^3 **12.** 3 yd^3 ● 3 m^3

13. BAKING The bread loaf pan shown is filled to a height of 2 inches with banana bread batter. How much more batter could the pan hold before it overflowed?

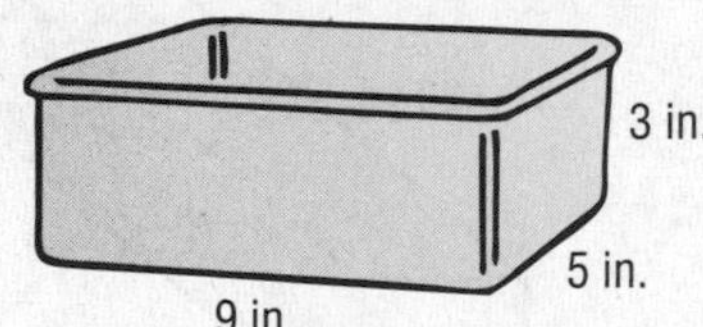

Copyright © Glencoe/McGraw-Hill, a division of The McGraw-Hill Companies, Inc.

NAME ______________________ DATE ____________ PERIOD _____

10-7 Study Guide and Intervention

Surface Area of Rectangular Prisms

The **surface area** S of a rectangular prism with length ℓ, width w, and height h is the sum of the areas of the faces.

Symbols $S = 2\ell w + 2\ell h + 2wh$

Model

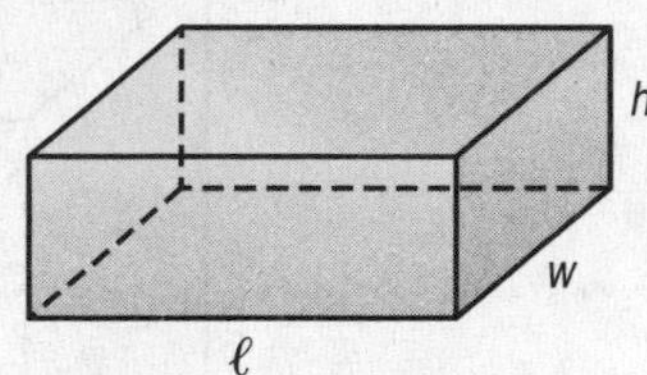

Example **Find the surface area of the rectangular prism.**

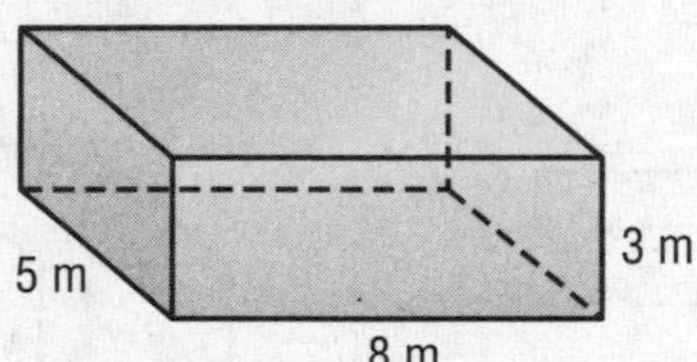

Find the area of each face.

top and bottom
$2(\ell w) = 2(8 \times 5) = 80$

front and back
$2(\ell h) = 2(8 \times 3) = 48$

two sides
$2(wh) = 2(5 \times 3) = 30$

Add to find the surface area. The surface area is 80 + 48 + 30 or 158 square meters.

Exercises

Find the surface area of each rectangular prism.

1.

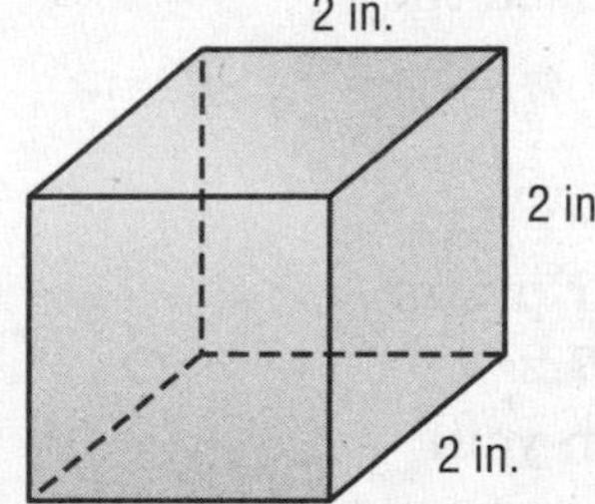

2.

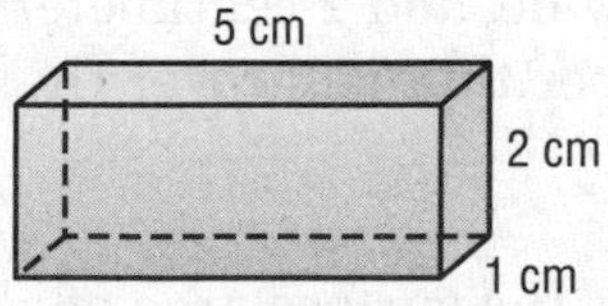

3.

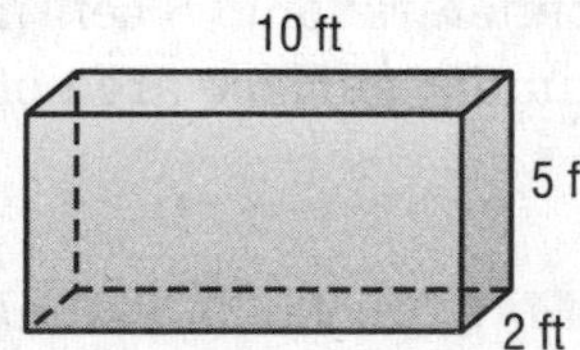

4.

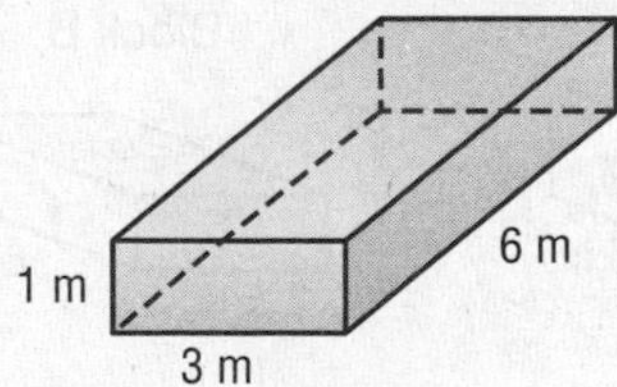

5.

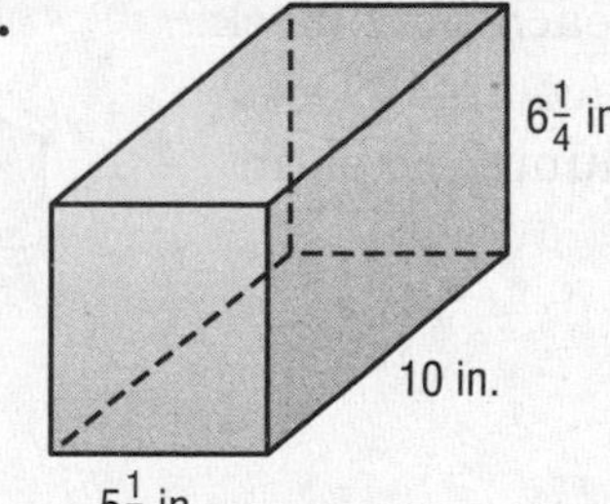

6.

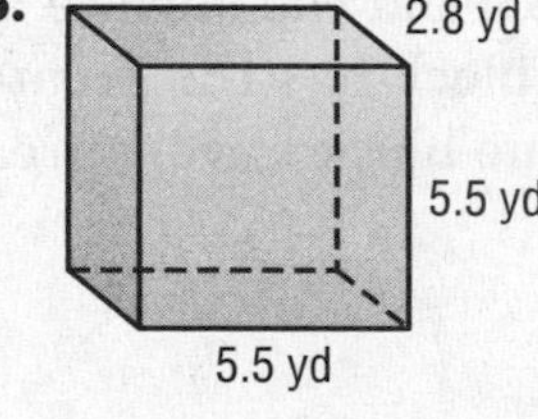

Lesson 10–7

Copyright © Glencoe/McGraw-Hill, a division of The McGraw-Hill Companies, Inc.

NAME ______________________ DATE ____________ PERIOD ____

10-7 Practice

Surface Area of Rectangular Prisms

Find the surface area of each rectangular prism.

1.

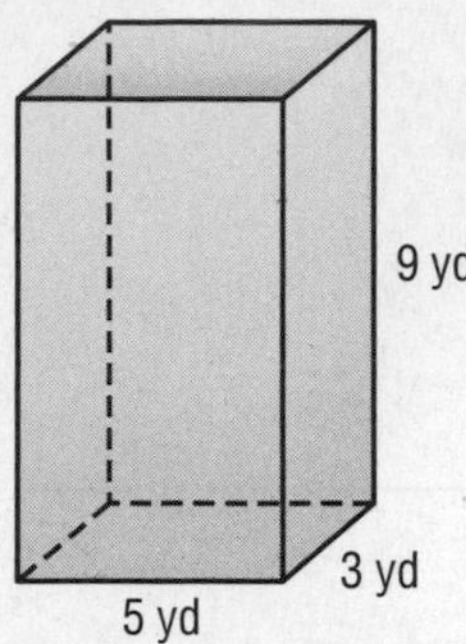

2.

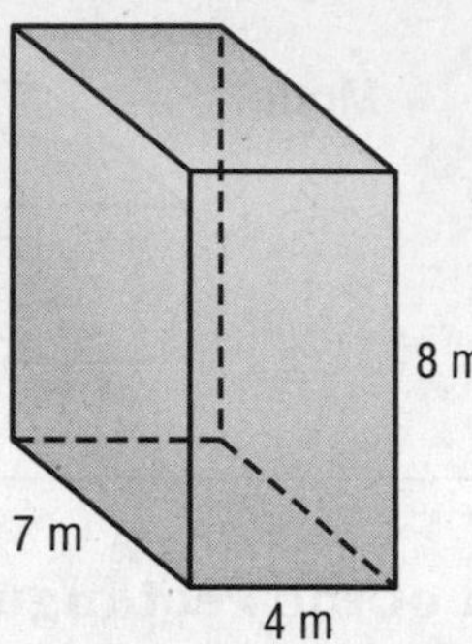

3.

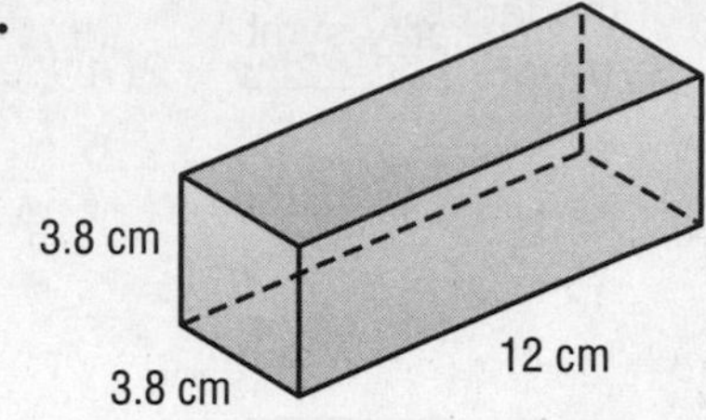

4.

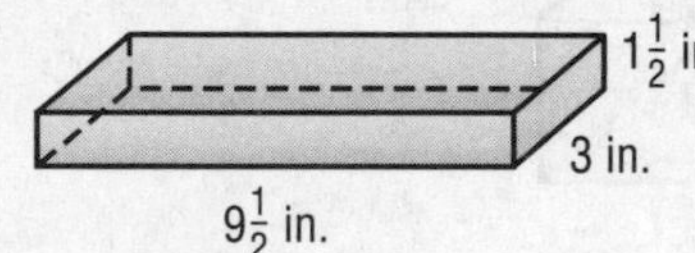

5.

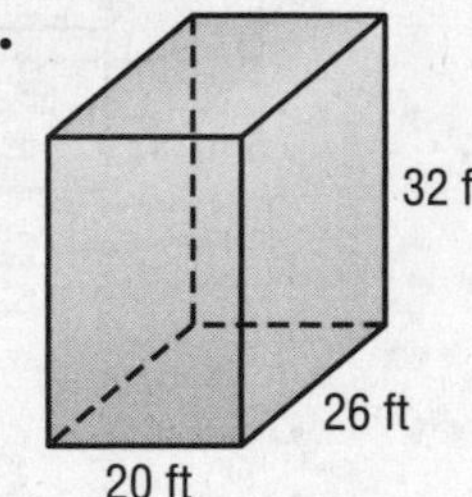

6.

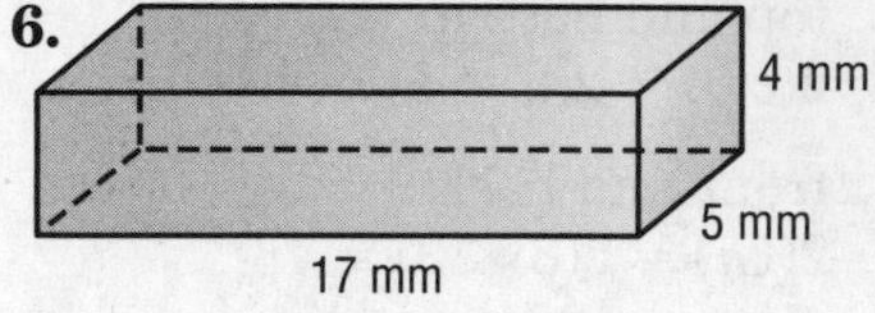

7. GIFTS Eric is covering a calculator with gift wrap. The calculator is 15 centimeters long, 8 centimeters wide, and 2 centimeters high. What is the minimum surface area of the paper that will cover the calculator?

8. ESTIMATION Alicia estimates that the surface area of a rectangular prism with a length of 11 meters, a width of 5.6 meters, and a height of 7.2 meters is about 334 cubic feet. Is her estimate reasonable? Explain your reasoning.

9. BLOCKS Find the surface area of each play block. Which block has the greater surface area? Does the same block have a greater volume? Explain.

Block A

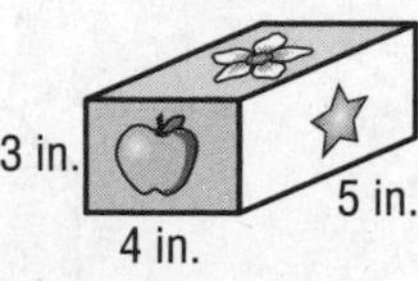

Block B

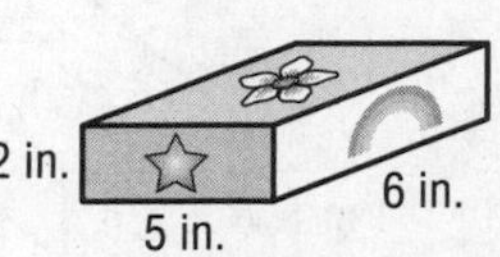

Copyright © Glencoe/McGraw-Hill, a division of The McGraw-Hill Companies, Inc.

NAME ______________________________ DATE ____________ PERIOD _____

11-1 Study Guide and Intervention

Ordering Integers

The inequality symbol '>' means *is greater than.*

The inequality symbol '<' means *is less than.*

Example 1 **Replace ● with < or > to make the statement 4 ● −5 true.**

Graph 4 and −5 on a number line. Then compare.

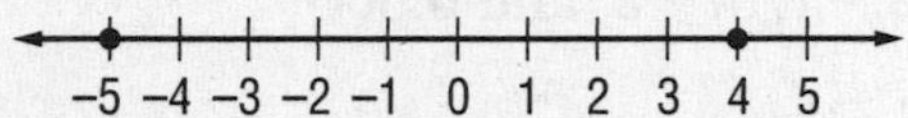

Since 4 is to the right of −5, 4 > −5 is a true statement.

Example 2 **Order the integers 1, −2, and 3 from least to greatest.**

Graph each integer on a number line. Then compare.

The order from least to greatest is −2, 1, and 3.

Exercises

Replace each ● with < or > to make a true statement.

1. −2 ● 0

2. 3 ● −3

3. −9 ● 8

4. −8 ● −3

5. 11 ● 3

6. −2 ● 10

Order each set of integers from least to greatest.

7. −2, 3, 0, −1, 1

8. 3, −3, −2, 1, −1

9. 5, −7, −2, 1, 9

10. −2, 1, 5, −5, 0

Copyright © Glencoe/McGraw-Hill, a division of The McGraw-Hill Companies, Inc.

NAME ________________ DATE ________ PERIOD ______

11-1 Practice

Ordering Integers

Replace each ● with < or > to make a true sentence.

1. 18 ● 23 **2.** −9 ● −1 **3.** −3 ● −5 **4.** 8 ● −2

5. 6 ● −3 **6.** 0 ● 8 **7.** 6 ● −7 **8.** −23 ● −16

Order each set of integers from least to greatest.

9. 10, −5, 3 16, −1, 0, and 1

10. −2.5, 4, 23, −1, 5, −3, and 0.66

11. 1, −2.5, 0.75, 3, and −0.75

12. 63, −34, 36, −27, −13, and 12

Order each set of integers from greatest to least.

13. 8, 43, −25, 12, −14, and 3

14. −8, 32, 55, −32, −19, and −3

15. −100, −89, −124, −69, and −52

16. 6, 17, −20, 15, −19, and 26

ROLLER COASTERS The table shows how several roller coasters compare to the Mantis. Refer to the table to answer Exercises 17–20.

Roller Coaster	Lift Heights (ft)	Vertical Drop (ft)
Gemini	−20	−19
Magnum XL-200	60	58
Top Thrill Dragster	275	263
Mantis	0	0
Millenium Force	165	163
Mean Streak	16	18
Raptor	−8	−18

Source: Cedar Point

17. Which roller coaster has the greatest lift height?

18. What is the median lift height for the roller coasters listed? Round to the nearest tenth.

19. Arrange the given roller coasters from least to greatest lift height.

20. What is the median of the data for vertical drop?

Copyright © Glencoe/McGraw-Hill, a division of The McGraw-Hill Companies, Inc.

NAME ______________________________ DATE ______________ PERIOD _____

11-2 Study Guide and Intervention

Adding Integers

- The sum of two positive integers is always positive.
- The sum of two negative integers is always negative.
- The sum of a positive integer and a negative integer is sometimes positive, sometimes negative, and sometimes zero.

Example 1 **Find −3 + (−2).**

Method 1 Use counters.

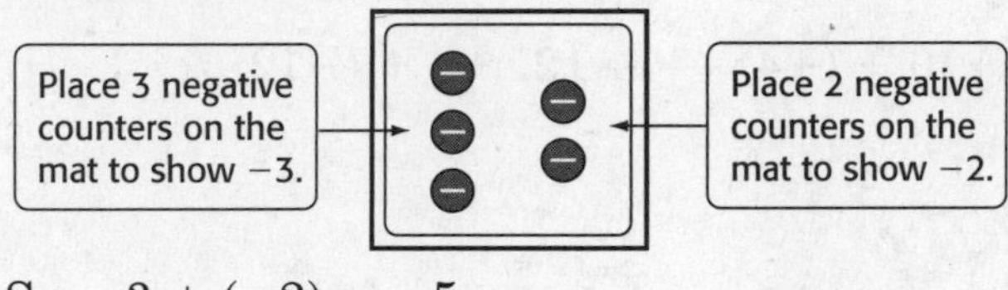

So, −3 + (−2) = −5.

Method 2 Use a number line.

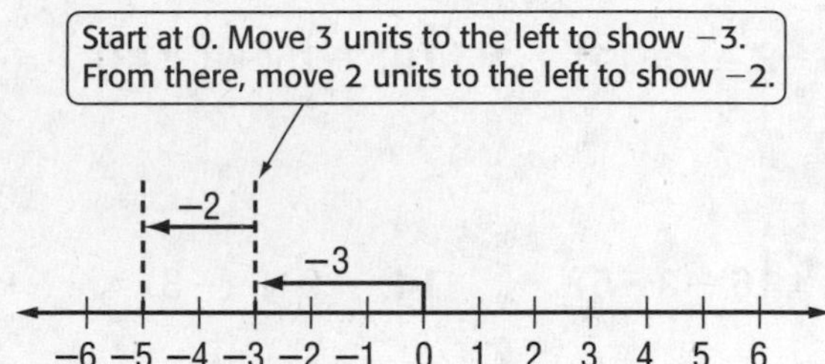

Example 2 **Find 4 + (−1).**

Method 1 Use counters.

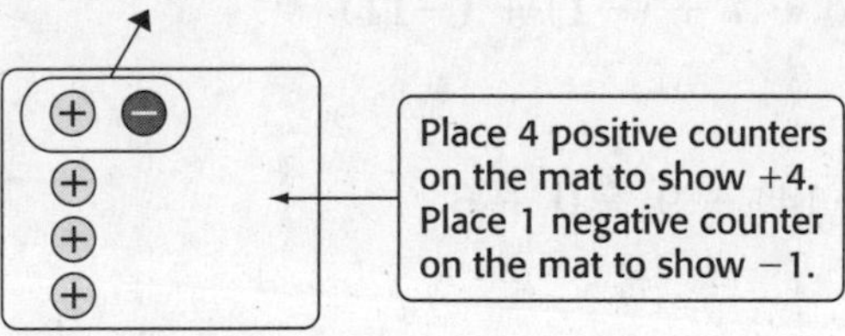

So, 4 + (−1) = 3.

Method 2 Use a number line.

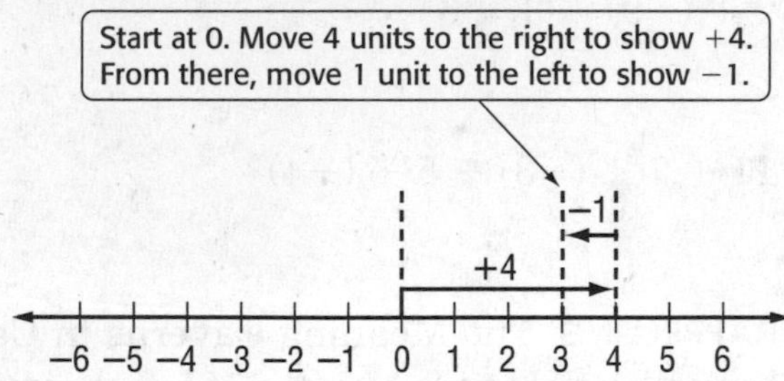

Exercises

Add. Use counters or a number line if necessary.

1. 3 + (−6) **2.** −9 + 8 **3.** −4 + 7

4. 6 + (−6) **5.** −8 + (−2) **6.** 2 + (−5)

7. 6 + (−12) **8.** −6 + (−5) **9.** 4 + (−3)

10. −12 + 5 **11.** −4 + 10 **12.** −3 + (−5)

Lesson 11–2

Copyright © Glencoe/McGraw-Hill, a division of The McGraw-Hill Companies, Inc.

NAME ______________________ DATE ____________ PERIOD ____

11-2 Practice

Adding Integers

Add. Use counters or a number line if necessary.

1. +8 + (+4) **2.** −10 + (+7) **3.** −2 + (−10) **4.** +9 + (−1)

5. −6 + (−5) **6.** +8 + (+9) **7.** +5 + (−3) **8.** −4 + (−9)

9. −2 + (+14) **10.** −15 + (+13) **11.** +10 + (+4) **12.** +8 + (−12)

13. +16 + (−5) **14.** +9 + (−3) **15.** −3 + (−8) **16.** −1 + (+1)

Add.

17. 2 + (−9) + 3 + 6 **18.** 3 + (−8) + 7 + (−1) + (−11)

19. 11 + 7 + (−3) + 5 + (−4) **20.** −2 + (−14) + 9 + 0 + 6

21. RAPPELLING The Moaning Caverns in California are 410 feet deep. A rappeller descends by rope 165 feet into the main cavern. How much deeper can the rappeller go into the cavern?

22. SEWING Keisha discovered a mistake in her cross-stitch project after she had completed a row. To remove the mistake she had to pull out 72 stitches. She then sewed 39 stitches before having to change to a new thread color. If her starting point is zero, at what point is she in the row now?

23. Which expression is represented by the number line below?

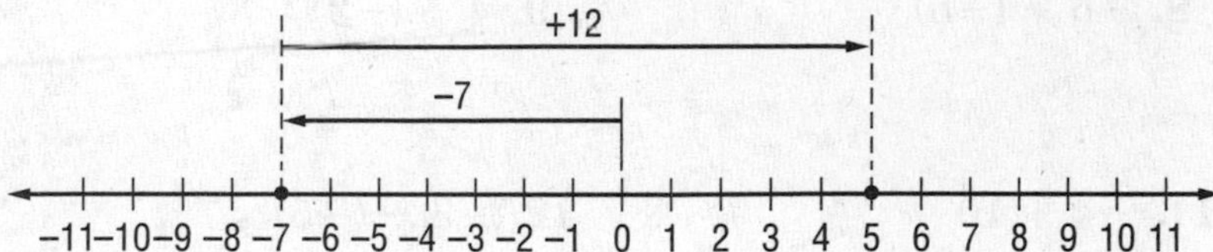

Copyright © Glencoe/McGraw-Hill, a division of The McGraw-Hill Companies, Inc.

NAME ______________ DATE ________ PERIOD ____

11-3 Study Guide and Intervention

Subtracting Integers

To subtract an integer, add its opposite.

Example 1 **Find −4 − (−3).**

Method 1 Use counters.

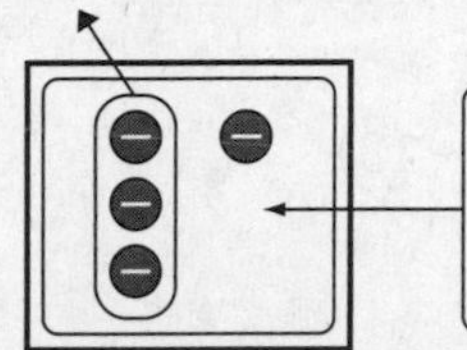

Place 4 negative counters on the mat to show −4. Remove 3 negative counters to show subtracting −3.

So, −4 − (−3) = −1.

Method 2 Use the rule.

$-4 - (-3) = -4 + 3$	To subtract −3, add 3.
$= -1$	Simplify.

Example 2 **Find −3 − 1.**

Method 1 Use counters.

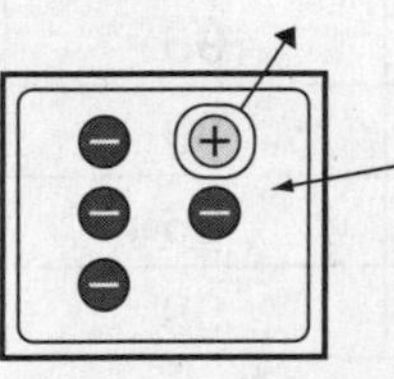

Place 3 negative counters on the mat to show −3. To subtract +1, you must remove 1 positive counter. But there are no positive counters on the mat. You must add 1 zero pair to the mat. The value of the mat does not change. Then you can remove 1 positive counter.

The difference of −3 and 1 is −4.

So, −3 −1 = −4.

Method 2 Use the rule.

$-3 - 1 = -3 + (-1)$	To subtract 1, add −1.
$= -4$	Simplify.

Exercises

Subtract. Use counters if necessary.

1. +8 − 5 **2.** −4 − 2 **3.** 7 − (−5)

4. −3 − (−5) **5.** 6 − (−10) **6.** −8 − (−4)

7. −1 − 4 **8.** 2 − (−2) **9.** −5 − (−1)

10. 7 − 2 **11.** −9 − (−9) **12.** 6 − (−2)

13. −8 − (−14) **14.** −2 − 9 **15.** 5 − 15

Lesson 11–3

Copyright © Glencoe/McGraw-Hill, a division of The McGraw-Hill Companies, Inc.

NAME ______________________ DATE ____________ PERIOD ____

11-3 Practice

Subtracting Integers

Subtract. Use counters if necessary.

1. 12 − 9 **2.** 11 − 13 **3.** −6 − 15 **4.** 8 − 4

5. −8 − (−15) **6.** −8 − (−5) **7.** 10 − (−12) **8.** −1 − 6

9. 5 − (−5) **10.** −7 − (−13) **11.** −17 − (−19) **12.** 3 − (−13)

13. −3 − 9 **14.** 14 − (−4) **15.** 0 − (−8) **16.** −13 − (−12)

17. The table at the right shows the results of two consecutive Biology tests for James, Mazen, Mia, and Shameeka. What is the test differential for each student?

Biology Test Results		
Student	**Test 1**	**Test 2**
James	84	96
Mazen	98	89
Mia	70	86
Shameeka	100	98

18. **ALGEBRA** Evaluate $c - d$ if $c = 4$ and $d = 9$.

19. The blue whale can dive as deep as 1,640 feet. A blue whale is at 600 feet below sea level and rises 370 feet to feed. It then dives 90 feet. Where is it?

20. **SWIMMING** Sara swims at the community center every day. One week she swam a total of 13.5 hours. Complete the table.

Day	Number of Hours
Monday	1.5
Tuesday	2.0
Wednesday	1.5
Thursday	1.5
Friday	2.0
Saturday	
Sunday	2.0

Copyright © Glencoe/McGraw-Hill, a division of The McGraw-Hill Companies, Inc.

NAME ______________________________ DATE ______________ PERIOD _____

11-4 Study Guide and Intervention

Multiplying Integers

- The product of two integers with different signs is negative.
- The product of two integers with the same sign is positive.

Examples **Multiply.**

1 $2 \times (-1)$

$2 \times (-1) = -2$ The integers have different signs. The product is negative.

2 -4×3

$-4 \times 3 = -12$ The integers have different signs. The product is negative.

3 3×5

$3 \times 5 = 15$ The integers have the same sign. The product is positive.

4 $-2 \times (-4)$

$-2 \times (-4) = 8$ The integers have the same sign. The product is positive.

Exercises

Multiply.

1. $3 \times (-3)$ **2.** $-5 \times (-2)$ **3.** $-8 \times (-1)$

4. -2×8 **5.** 4×-3 **6.** $-3 \times (-2)$

7. $5 \times (-4)$ **8.** $-10 \times (-4)$ **9.** -3×6

10. $-3 \times (-10)$ **11.** $6 \times (-4)$ **12.** $-7 \times (-7)$

Lesson 11–4

Copyright © Glencoe/McGraw-Hill, a division of The McGraw-Hill Companies, Inc.

NAME ______________________________ DATE ____________ PERIOD _____

11-4 Practice

Multiplying Integers

Multiply.

1. -2×15 **2.** $-4 \times (-11)$ **3.** $-3 \times (-3)$ **4.** $7(2)$

5. $6(-8)$ **6.** 13×8 **7.** $15(-6)$ **8.** -12×3

9. $-10(-4)$ **10.** $-1(-7)$ **11.** 8×3 **12.** $-6 \times (-4)$

13. 13×7 **14.** $2 \times (-6)$ **15.** -9×9 **16.** $-3(-14)$

17. $9(-3 - 8)$ **18.** $-7(4)(-5)$ **19.** $-2(6 + (-7))$ **20.** $7(-3 + 3)$

21. $-2(8 + (-6))$ **22.** $3(-5)(2)$ **23.** $4(-2 + 9)$ **24.** $-3(-4 - 4)$

25. **PATTERNS** Find the next two numbers in the pattern. Then describe the pattern.

$$8, -24, 72, -216, \ldots$$

26. **ALGEBRA** Find the value of mn if $m = -7$ and $n = -12$.

27. **CONSTRUCTION** The arm and torch of the Statue of Liberty were completed for the International Centennial Exhibition in Philadelphia in 1876. It took 20 men working 10 hours a day, 7 days a week, to complete it for the exhibition. What was the total number of hours worked in a week?

28. **EXERCISE** After finishing her workout, Felicia's heart rate decreased by 2 beats per minute for each of the next 5 minutes. Write an integer to represent the change in her heart rate at the end of 5 minutes.

Copyright © Glencoe/McGraw-Hill, a division of The McGraw-Hill Companies, Inc.

11-5 Study Guide and Intervention

Problem-Solving Investigation: Work Backward

When solving problems, one strategy that is helpful is to *work backward.* Sometimes you can use information in the problem to work backwards to find what you are looking for, or the answer to the problem.

You can use the *work backward* strategy, along with the following four-step problem solving plan to solve a problem.

1 Understand – Read and get a general understanding of the problem.

2 Plan – Make a plan to solve the problem and estimate the solution.

3 Solve – Use your plan to solve the problem.

4 Check – Check the reasonableness of your solution.

Example TIME **Meagan is meeting her friends at the library at 6:30 P.M. Before her mom takes her to the library, they are going to stop by her grandma's house to drop something off. It takes 15 minutes to get from her house to her grandma's house and they will stay and visit for 30 minutes. If it takes 5 minutes to get from her grandma's house to the library, what time should Meagan and her mom leave their house?**

Understand We know the time Meagan is meeting her friends at the library. We need to find what time Meagan and her mom should leave their house.

Plan To find the time they should leave, start with the 6:30 P.M. and first subtract 5 minutes for the time it takes to get from her grandma's house to the library.

Solve
Time from grandma's to library: 6:30 P.M. – 5 minutes = 6:25 P.M.
Time visiting with grandma: 6:25 P.M. – 30 minutes = 5:55 P.M.
Time from home to grandma's: 5:55 P.M. – 15 minutes = 5:40 P.M.

Meagan and her mom should leave their house at 5:40 P.M.

Check Add up all the times, 15 min + 30 min + 5 min = 50 min. When you subtract 50 minutes from 6:30, the result is 5:40, so the answer is correct.

Exercise

NUMBER SENSE A number is divided by 3. Next, 7 is added to the quotient. Then, 10 is subtracted from the sum. If the result is 5, what is the number?

Copyright © Glencoe/McGraw-Hill, a division of The McGraw-Hill Companies, Inc.

NAME ______________________________ DATE ____________ PERIOD ______

11-5 Practice

Problem-Solving Investigation: Work Backward

Mixed Problem Solving

Work backward to solve Exercises 1 and 2.

1. **NUMBER SENSE** A number is multiplied by 4. Next, 3 is added to the product, and then 11 is subtracted. If the result is 24, what is the number?

2. Ichiko has guitar practice at 5:00 P.M. on Wednesday. It takes 20 minute for him to get to his lesson from school. He spends an hour in the science lab before leaving. If it takes 10 minutes to get ready for the lab, what time does his last class end?

Use any strategy to solve Exercises 3–6. Some strategies are shown below.

Problem-Solving Strategies
• Act it out.
• Make a table.
• Choose the method of computation.

3. **GEOGRAPHY** North America has an area of 21,393,762 square kilometers. South America has an area of 17,522,371 square kilometers. What is the combined area of these two continents?

4. **FLIGHT SCHOOL** The list shows how many times each of 20 students practiced with a piloting simulator at a flight training school one day.

   9 11 12 9 6 12 10 8 13 14
   8 9 13 11 10 8 12 9 10 8

   Make a frequency table to find how many more students practiced with the simulator 9–11 times than 12–14 times.

5. **FOOD** The total cost for a take-out lunch was $20. If four friends share the cost equally, how much will each friend pay?

6. **MONEY** Mai had $210 in her checking account at the beginning of the month. She wrote checks for $32 and $9.59. At the end of the month, the bank credited her account with $0.84 interest. How much money did Mai have in the account then?

Copyright © Glencoe/McGraw-Hill, a division of The McGraw-Hill Companies, Inc.

NAME ______________________ DATE __________ PERIOD _____

11-6 Study Guide and Intervention

Dividing Integers

- The quotient of two integers with different signs is negative.
- The quotient of two integers with the same sign is positive.

Example 1 **Use counters to find $-6 \div 2$.**

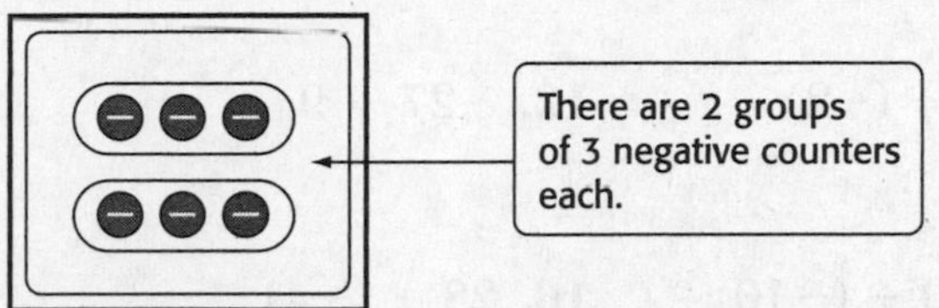

So, $-6 \div 2 = -3$.

Examples **Divide.**

2 $\mathbf{10 \div (-5)}$

Since $-5 \times (-2) = 10$, it follows that $10 \div (-5) = -2$.

3 $\mathbf{-12 \div (-3)}$

Since $-3 \times 4 = -12$, it follows that $-12 \div (-3) = 4$.

Exercises

Divide.

1. $4 \div (-2)$ **2.** $-9 \div (-3)$ **3.** $-8 \div 2$

4. $-21 \div 7$ **5.** $30 \div (-5)$ **6.** $-24 \div 4$

7. $-36 \div 6$ **8.** $-45 \div (-5)$ **9.** $-81 \div 9$

10. $-3 \div (-3)$ **11.** $70 \div (-7)$ **12.** $-64 \div (-8)$

13. ALGEBRA Find the value of $a \div b$ if $a = -18$ and $b = 6$.

14. ALGEBRA For what value of p is $p \div 5 = -7$ true?

Copyright © Glencoe/McGraw-Hill, a division of The McGraw-Hill Companies, Inc.

NAME ______________________________ DATE ____________ PERIOD _____

11-6 Practice

Dividing Integers

Divide.

1. $33 \div (-3)$

2. $-84 \div (-6)$

3. $-26 \div 13$

4. $92 \div (-23)$

5. $-96 \div 4$

6. $36 \div (-6)$

7. $76 \div 4$

8. $-12 \div (-6)$

9. $-30 \div (-5)$

10. $-42 \div 7$

11. $18 \div (-2)$

12. $-27 \div 9$

13. $69 \div 23$

14. $52 \div 13$

15. $-40 \div (-10)$

16. $28 \div (-4)$

17. $\frac{-8 - 7}{-5}$

18. $\frac{5 - (-4) + (-9 + 6)}{-6}$

19. $\frac{(21 \div 3) \times 8}{-4}$

20. $\frac{(-3 + (-2)) \times (-6 + 1)}{5}$

21. **MILKING** It takes 20 minutes for a cow to be milked by a milking machine. How many cows can be milked in 6 hours?

22. **ALGEBRA** What is the value of $s \div t$ if $s = -18$ and $t = -6$?

23. **TESTING** Thi wants to find the average of her last four math tests. She scored 96 on her first test. Use the table to find her average score for the four tests.

Thi's Tests	
Test 1	0
Test 2	−13
Test 3	−5
Test 4	3

24. **GASOLINE** The price of a gallon of gasoline increased by 5 cents one week, decreased by 3 cents each of the next two weeks, and increased by 7 cents the fourth week. Find the average change in the price of gasoline for the 4 weeks.

Copyright © Glencoe/McGraw-Hill, a division of The McGraw-Hill Companies, Inc.

NAME ______________________ DATE __________ PERIOD _____

11-7 Study Guide and Intervention

The Coordinate Plane

The *x*-axis and *y*-axis separate the coordinate system into four regions called **quadrants**.

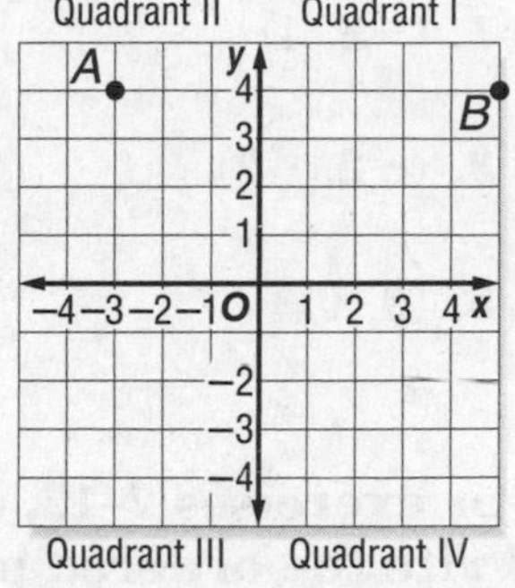

Example 1 **Identify the ordered pair that names point *A*.**

Step 1 Move left on the *x*-axis to find the *x*-coordinate of point *A*, which is −3.

Step 2 Move up the *y*-axis to find the *y*-coordinate, which is 4.

Point ***A*** is named by (−3, 4).

Example 2 **Graph point *B* at (5, 4).**

Use the coordinate plane shown above. Start at 0. The *x*-coordinate is 5, so move 5 units to the right.

Since the *y*-coordinate is 4, move 4 units up.

Draw a dot. Label the dot *B*.
See grid at the top of the page.

Exercises

Use the coordinate plane at the right. Write the ordered pair that names each point.

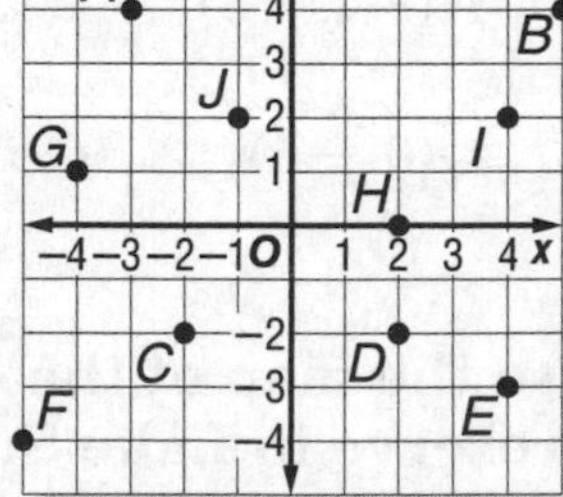

1. *C*

2. *D*

3. *E*

4. *F*

5. *G*

6. *H*

7. *I*

8. *J*

Graph and label each point using the coordinate plane at the right.

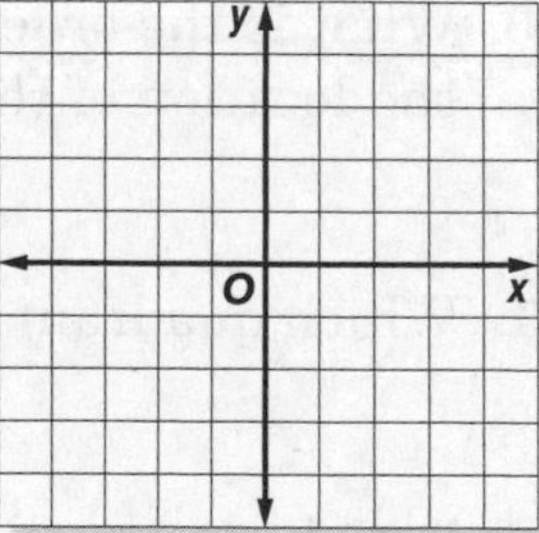

9. $A(-5, 2)$

10. $M(-4, -5)$

11. $G(-1, -4)$

12. $L(0, -1)$

13. $N(1, 5)$

14. $I(2, -1)$

Copyright © Glencoe/McGraw-Hill, a division of The McGraw-Hill Companies, Inc.

NAME ______________________ DATE __________ PERIOD _____

11-7 Practice

The Coordinate Plane

Use the coordinate plane at the right for Exercises 1–6. Identify the point for each ordered pair.

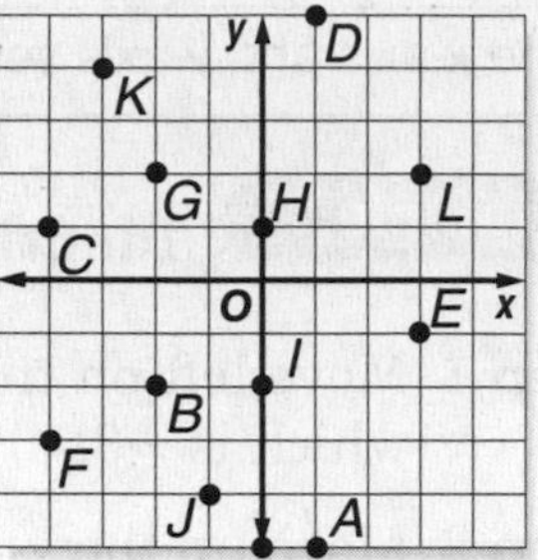

1. (−3, 4)
2. (−4, −3)
3. (−2, −2)
4. (3, −1)
5. (0, 1)
6. (−1, −4)

For exercises 7–12, use the coordinate plane above. Write the ordered pair that names each point. Then identify the quadrant where each point is located.

7. *C*
8. *L*
9. *D*
10. *A*
11. *G*
12. *I*

Graph and label each point on the coordinate plane at the right.

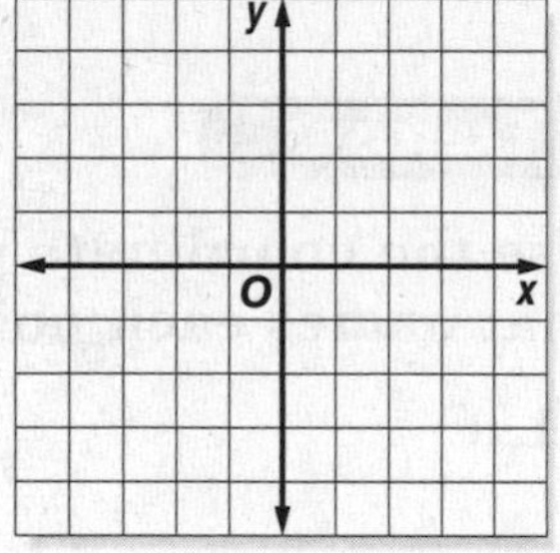

13. $L(-2, 0)$
14. $M(5, 2)$
15. $N(-4, -3)$
16. $P(1, -1)$
17. $Q(0, -4)$
18. $R(3, -3)$

Use the map of the Alger Underwater Preserve in Lake Superior to answer the following questions.

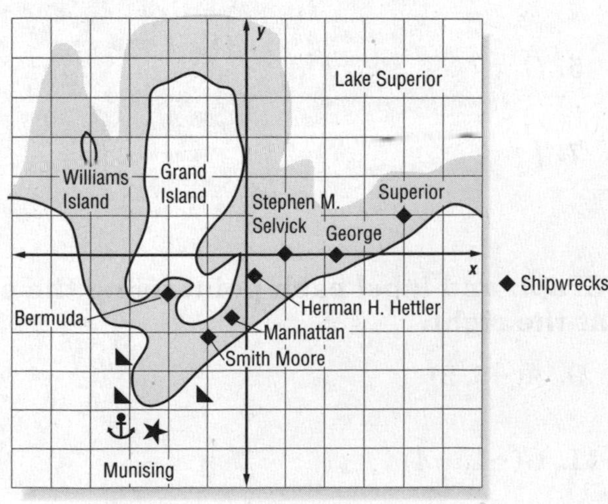

19. In which quadrant is the Stephen M. Selvick located?

20. What is the ordered pair that represents the location of the Bermuda? the Superior?

21. Which quadrant contains Williams Island?

22. Which shipwreck is closest to the origin?

Copyright © Glencoe/McGraw-Hill, a division of The McGraw-Hill Companies, Inc.

NAME ______________________ DATE ____________ PERIOD _____

11-8 Study Guide and Intervention

Translations

- A transformation is a movement of a geometric figure.
- The resulting figure is called the image.
- A translation is the sliding of a figure without turning it.
- A translation does not change the size or shape of a figure.

Example 1 **Translate triangle *ABC* 5 units to the right.**

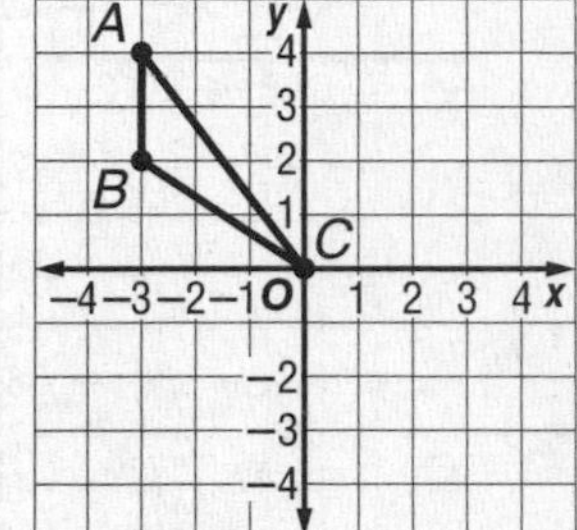

Step 1 Move each vertex of the triangle 5 units right. Label the new vertices *A'*, *B'*, *C'*.

Step 2 Connect the new vertices to draw the triangle. The coordinates of the vertices of the new triangle are *A'*(2, 4), *B'*(2, 2), and *C'*(5, 0).

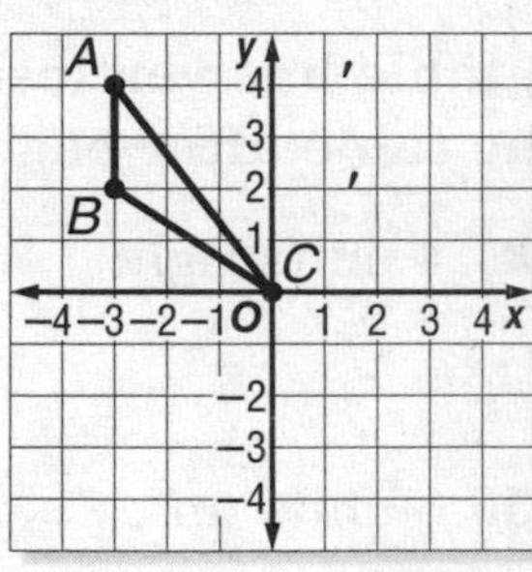

Example 2 A placemat on a table has vertices at (0, 0), (3, 0), (3, 4), and (0, 4). Find the vertices of the placemat after a translation of 4 units right and 2 units up.

Vertex	$(x + 4, y + 2)$	New vertex
(0, 0)	(0 + 4, 0 + 2)	(4, 2)
(3, 0)	(3 + 4, 0 + 2)	(7, 2)
(3, 4)	(3 + 4, 4 + 2)	(7, 6)
(0, 4)	(0 + 4, 4 + 2)	(4, 6)

Exercises

Find the coordinates of the image of (2, 4), (1, 5), (1, −3), and (3, −4) under each transformation.

1. 2 units right

2. 4 units down

3. 3 units left and 4 units down

4. 5 units right and 3 units up

Lesson 11–8

Copyright © Glencoe/McGraw-Hill, a division of The McGraw-Hill Companies, Inc.

NAME ______________________ DATE ________ PERIOD ______

11-8 Practice

Translations

1. Translate *LMN* 5 units down. Graph triangle *L′M′N′*.

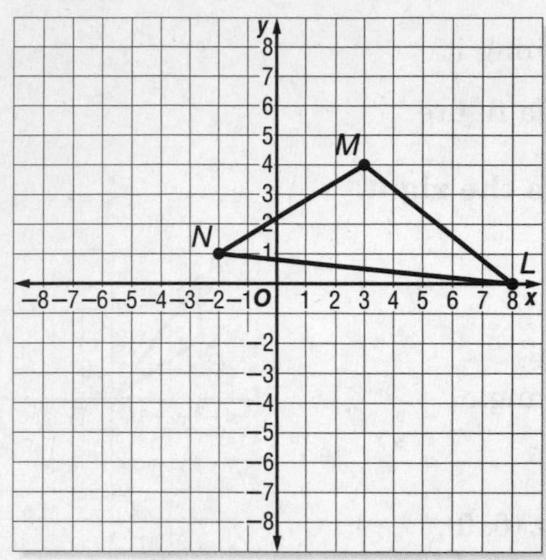

2. Translate *TRI* 2 units left and 3 units up. Graph *T′R′I′*.

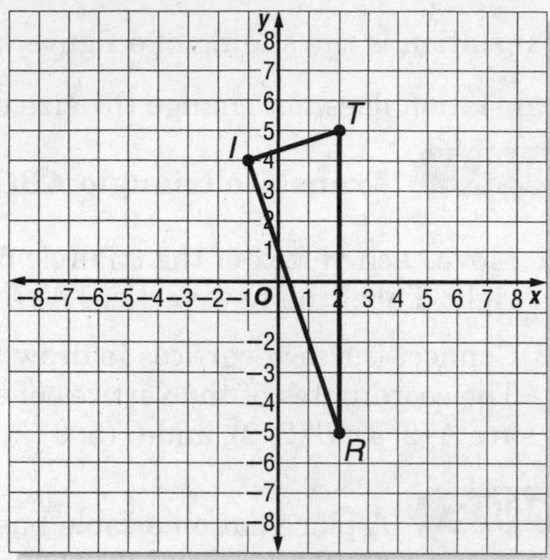

A table has vertices of (0, 3), (6, 2), (0, 8), and (−2, 5) on a floor. Find the vertices of the table after each translation.

3. 4 units right

4. 2 units left

5. 6 units up

6. 9 units down

7. 1 unit left and 5 units up

8. 7 units right and 8 units up

9. 4 units left and 6 units down

10. 9 units right and 3 units down

11. 1 unit left and 9 units up

12. 5 units right and 7 units down

13. One of the vertices of a square is (3, 5). What is the ordered pair of the image after a translation of 3 units up, 5 units left and then 4 units down? What translation will give the same result?

Copyright © Glencoe/McGraw-Hill, a division of The McGraw-Hill Companies, Inc.

NAME ______________________ DATE ____________ PERIOD _____

11-9 Study Guide and Intervention

Reflections

- A reflection is the mirror image that is created when a figure is flipped over a line.
- A reflection is a type of geometric transformation.
- When reflecting over the x-axis, the y-coordinate changes to its opposite.
- When reflecting over the y-axis, the x-coordinate changes to its opposite.

Example 1 **Reflect triangle *ABC* over the x-axis.**

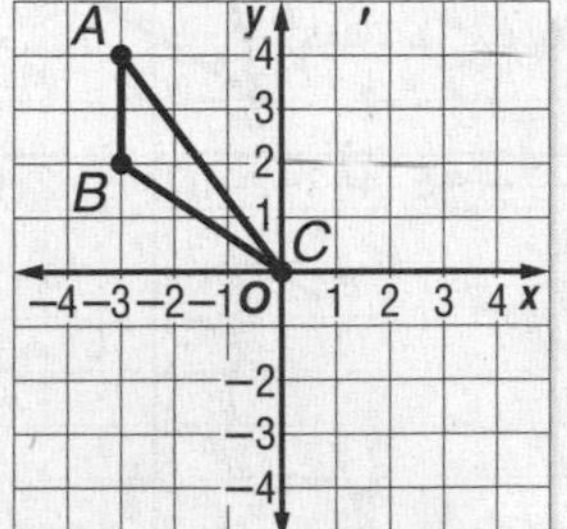

Step 1 Graph triangle *ABC* on a coordinate plane. Then count the number of units between each vertex and the x-axis.

A is 4 units from the axis.
B is 2 units from the axis.
C is 0 units from the axis.

Step 2 Make a point for each vertex the same distance away from the x-axis on the opposite side of the x-axis and connect the new points to form the image of the triangle. The new points are $A'(-3, -4)$, $B'(-2, -2)$, and $C'(0, 0)$.

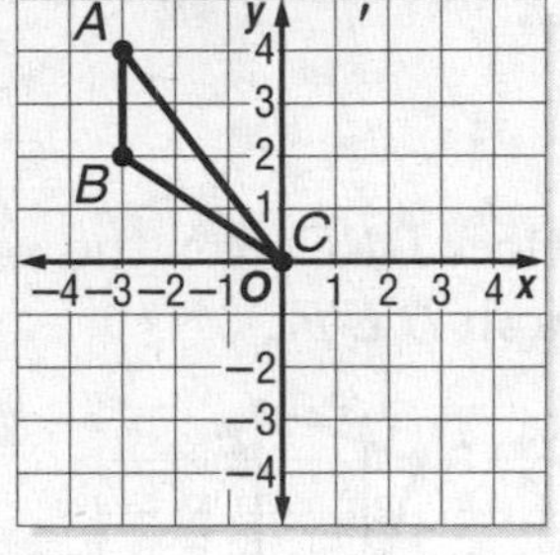

Exercises

Find the coordinates of the image of (2, 4), (1, 5), (1, −3) and (3, −4) under each transformation.

1. a reflection over the x-axis

2. a reflection over the y-axis.

Find the coordinates of the image of (−1, 1), (3, −2) and (0, 5) under each transformation.

3. a reflection over the x-axis

4. a reflection over the y-axis

Lesson 11–9

Copyright © Glencoe/McGraw-Hill, a division of The McGraw-Hill Companies, Inc.

NAME ______________________________ DATE __________ PERIOD ______

11-9 Practice

Reflections

1. Reflect *PQR* over the *x*-axis. Graph *P'Q'R'*.

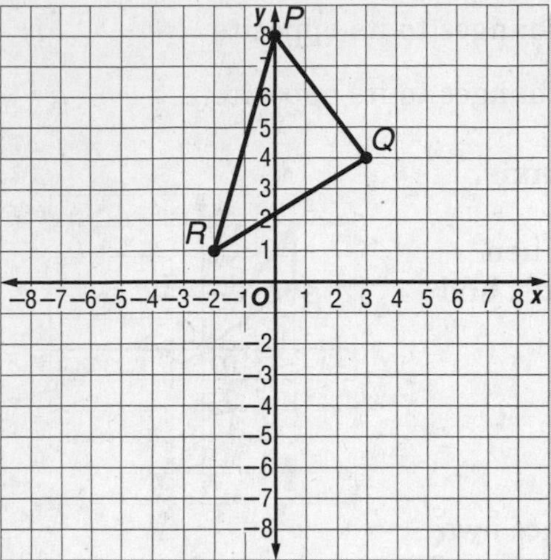

2. Reflect *PQR* over the *y*-axis. Graph *P'Q'R'*.

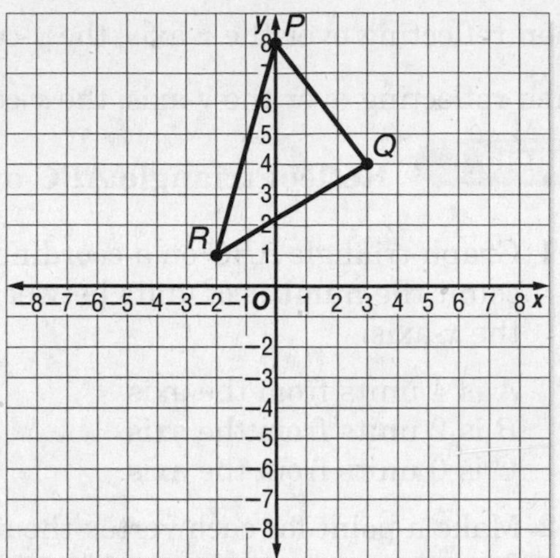

3. Reflect *DEF* over the *x*-axis. Graph *D'E'F'*.

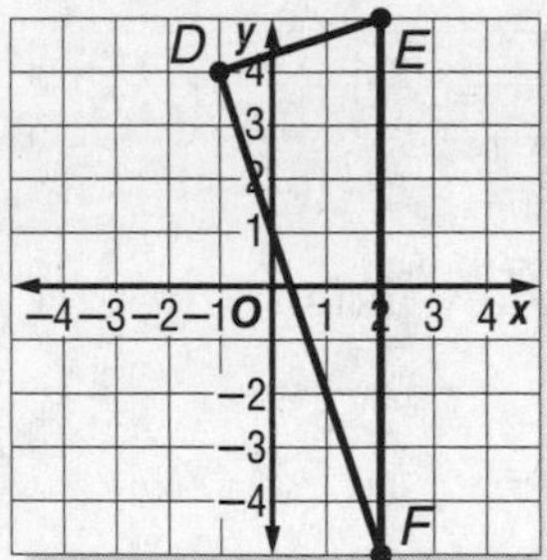

4. Reflect *DEF* over the *y*-axis. Graph *D'E'F'*.

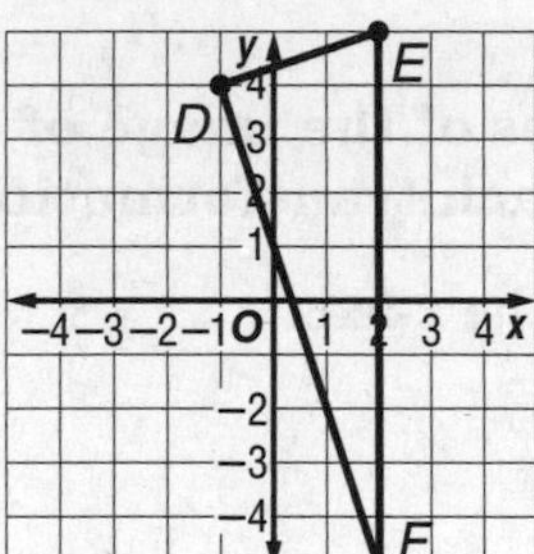

A table has vertices of (0, 3), (6, 2), (0, 8), and (−2, 5) on a floor. Find the vertices of the table after each transformation.

5. a reflection over the *x*-axis

6. a reflection over the *y*-axis.

A piece of artwork has vertices of (2, 5), (−1, 6) and (0, −7) on a wall. Find the vertices of the table after each transformation.

7. a reflection over the *x*-axis

8. a reflection over the *y*-axis

Copyright © Glencoe/McGraw-Hill, a division of The McGraw-Hill Companies, Inc.

NAME ______________________ DATE ____________ PERIOD ____

11-10 Study Guide and Intervention

Rotations

- A rotation occurs when a figure is rotated around a point.
- Another name for a rotation is a turn.
- In a rotation clockwise of 90° about the origin, the point (x, y) becomes $(y, -x)$.
- In a rotation clockwise of 180° about the origin, the point (x, y) becomes $(-x, -y)$.
- In a rotation clockwise of 270° about the origin, the point (x, y) becomes $(-y, x)$.

Example 1 **Rotate triangle *ABC* clockwise 180° about the origin.**

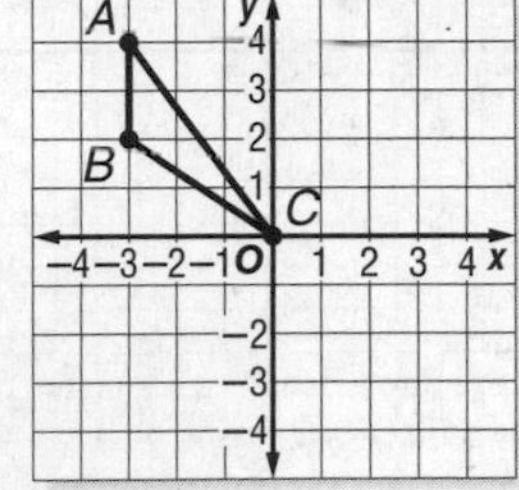

Step 1 Graph triangle *ABC* on a coordinate plane.

Step 2 Sketch segment *AO* connecting point *A* to the origin. Sketch another segment *A' O* so that the angle between point *A, O,* and *A'* measures 180° and the segment is congruent to *AO*.

Step 3 Repeat for point *B* (point *C* won't move since it is at the origin). Then connect the vertices to form triangle *A' B' C'*.

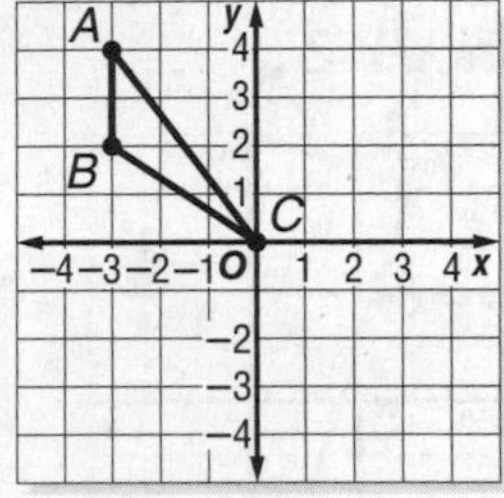

Exercises

Find the coordinates of the image of (2, 4), (1, 5), (1, −3) and (3, −4) under each transformation.

1. a rotation of 90° about the origin

2. a rotation of 270° about the origin

Determine whether each figure has rotational symmetry. Write *yes* or *no*. If yes, name the angle(s) of rotation.

3. X

2. Q

Lesson 11–10

Copyright © Glencoe/McGraw-Hill, a division of The McGraw-Hill Companies, Inc.

NAME ______________________ DATE __________ PERIOD ______

11-10 Practice
Rotations

1. Rotate ABC 90° about the origin. Graph $A'B'C'$.

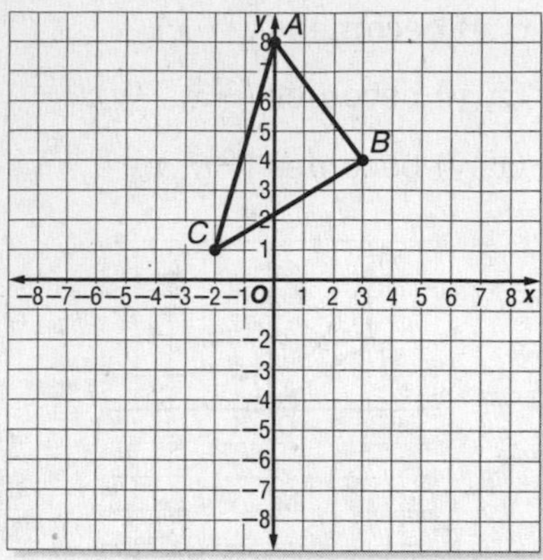

2. Rotate ABC 180° about the origin. Graph $A'B'C'$.

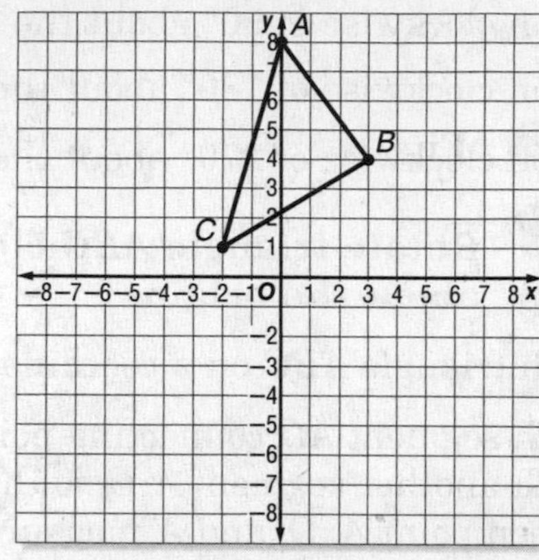

3. Rotate XYZ 270° about the origin. Graph $X'Y'Z'$.

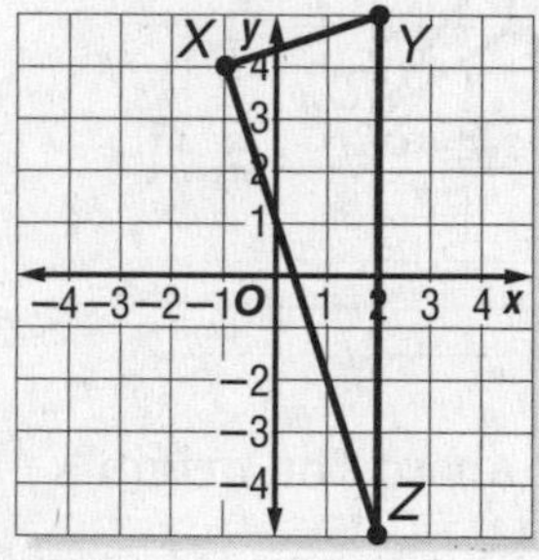

4. Rotate XYZ 180° about the origin. Graph $X'Y'Z'$.

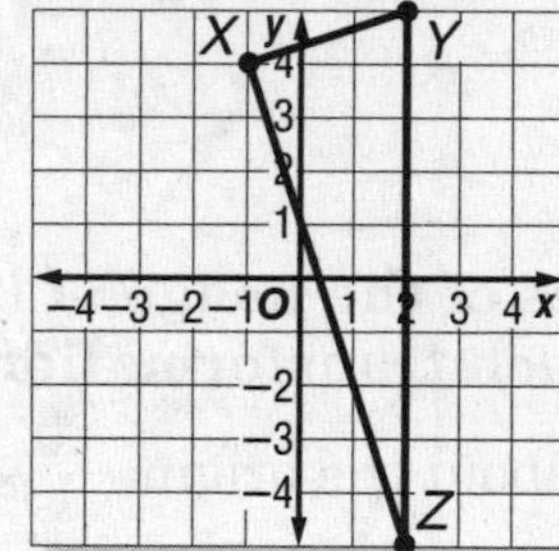

Determine whether each figure has rotational symmetry. Write *yes* or *no*. If yes, name the angle(s) of rotation.

5. N

6. H

7. 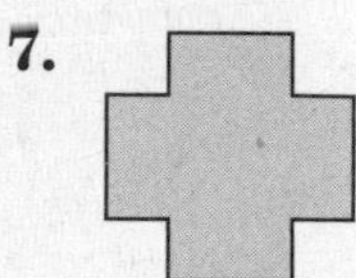

8. 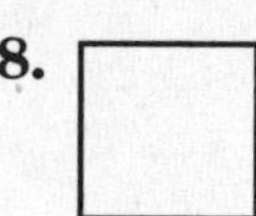

Copyright © Glencoe/McGraw-Hill, a division of The McGraw-Hill Companies, Inc.

12-1 Study Guide and Intervention

The Distributive Property

- To multiply a sum by a number, multiply each addend by the number outside the parentheses.
- $a(b + c) = ab + bc$
- $(b + c)a = ba + ca$

Example 1 **Find 6 × 38 mentally using the Distributive Property.**

$6 \times 38 = 6(30 + 8)$	Write 38 as 30 + 8.
$= 6(30) + 6(8)$	Distributive Property
$= 180 + 48$	Multiply mentally.
$= 228$	Add.

So, $6 \times 38 = 228$

Example 2 **Use the Distributive Property to rewrite $4(x + 3)$.**

$4(x + 3) = 4(x) + 4(3)$	Distributive Property
$= 4x + 12$	Multiply.

So, $4(x + 3)$ can be rewritten as $4x + 12$.

Exercises

Solve each problem mentally using the Distributive Property.

1. 4×82

2. 9×26

3. 12×44

4. 8×5.7

Use the Distributive Property to rewrite each algebraic expression.

5. $5(y + 4)$

6. $(7 + r)3$

7. $12(x + 5)$

8. $(b + 2)9$

9. $4(4 + a)$

10. $9(7 + v)$

Lesson 12–1

Copyright © Glencoe/McGraw-Hill, a division of The McGraw-Hill Companies, Inc.

NAME ______________________ DATE ____________ PERIOD _____

12-1 Practice

The Distributive Property

Solve each problem mentally using the Distributive Property.

1. 8×34 **2.** 5×47 **3.** 12×51

4. 8×53 **5.** 6×4.4 **6.** 7×2.9

Use the Distributive Property to rewrite each algebraic expression.

7. $6(n + 4)$ **8.** $(2 + r)15$ **9.** $8(s + 5)$

10. $(b + 8)3$ **11.** $5(6 + b)$ **12.** $9(3 + v)$

13. $(r - 7)7$ **14.** $12(4 - v)$ **15.** $11(3 - s)$

For Exercises 16–18, use the table that shows the prices of tickets and various food items at the movie theater.

Item	Price
Ticket	$8.50
Popcorn	$5.25
Soda	$4.00
Candy	$3.75
Nachos	$6.50

16. Four friends each bought a ticket and a bag of popcorn. How much total money did they spend?

17. How much money will the movie theater make if a birthday party of 12 kids each buys a box of candy and a soda but doesn't go see a movie?

18. How much more money will a person spend who buys three orders of nachos than a person who buys three bags of popcorn?

Copyright © Glencoe/McGraw-Hill, a division of The McGraw-Hill Companies, Inc.

NAME ______________________ DATE ____________ PERIOD _____

12-2 Study Guide and Intervention

Simplifying Algebraic Expressions

- **Commutative Property:** The order which numbers are added or multiplied does not change the sum or the product.
- $a + b = b + a$ or $a \cdot b = b \cdot a$.
- **Associative Property:** The way in which numbers are grouped does not change the sum or the product.
- $(a + b) + c = a + (b + c)$ or $(a \cdot b) \cdot c = a \cdot (b \cdot c)$
- **Like terms** contain the same variables. Ex: $2y$, y, and $7y$ are all like terms, but $4x$ is not.

Example 1 **Simplify the expression $16 + (v + 4)$.**

$16 + (v + 4)$	$= 16 + (4 + v)$	Commutative Property
	$= (16 + 4) + v$	Associative Property
	$= 20 + v$	Add.

So, $16 + (v + 4)$ in simplified form is $20 + v$.

Example 2 **Simplify the expression $3x + (6 + 2x)$.**

$3x + (6 + 2x)$	$= 3x + (2x + 6)$	Commutative Property
	$= (3x + 2x) + 6$	Associative Property
	$= 5x + 6$	Combine like terms.

So, $3x + (6 + 2x)$ in simplified form is $5x + 6$.

Exercises

Simplify each expression. Justify each step.

1. $5 + x + 3$

2. $6 + (x + 4)$

3. $(b + 10) + 15$

4. $8x + 5 + 2x$

5. $(12 + 2u) + 3$

6. $11p + 8 + 7p$

7. $9x + (4 + 3x)$

8. $(8 + 12x) + (2 + 7x)$

9. $5y + 4 + 7y$

Copyright © Glencoe/McGraw-Hill, a division of The McGraw-Hill Companies, Inc.

NAME ______________________ DATE __________ PERIOD _____

12-2 Practice

Simplifying Algebraic Expressions

Simplify each expression. Justify each step.

1. $(7 + x) + 7x$

2. $5 \cdot (4 \cdot x)$

3. $15 + (x + 9)$

4. $(6x + 21) + 14$

5. $3x + 2 + 11x$

6. $(x + 13) + 8$

7. $(12 + 2x) + 4$

8. $8 \cdot (x \cdot 4)$

9. $3(5x)$

10. $3x + (7x + 10)$

11. $5x + (2 + x)$

12. $4 \cdot x \cdot 10$

13. $(x \cdot 12) \cdot 3$

14. $14x + 9 + 6x$

15. $5x + (24 + 14x)$

ALGEBRA **For Exercises 16 through 21, translate each verbal expression into an algebraic expression. Then, simplify the expression.**

16. The sum of three and a number is added to twenty-four.

17. The product of six and a number is multiplied by nine.

18. The sum of 10 times a number and fifteen is added to eleven times the same number.

19. Two sets of the sum of a number and eight are added to five times the same number.

20. Three sets of a sum of a number and four are added to the sum of seven times the same number and thirteen.

21. Five friends went to a baseball game. Three of the friends each bought a ticket for x dollars and a soda for \$6.00. The other two friends each bought only tickets. Write and simplify an expression that represents the amount of money spent.

Copyright © Glencoe/McGraw-Hill, a division of The McGraw-Hill Companies, Inc.

NAME ______________________ DATE ____________ PERIOD _____

12-3 Study Guide and Intervention

Solving Addition Equations

Subtraction Property of Equality If you subtract the same number from each side of an equation, the two sides remain equal.

$$\begin{array}{r} 5 = 5 \\ -3 = -3 \\ \hline 2 = 2 \end{array}$$

Example 1 **Solve $x + 2 = 7$ using models.**

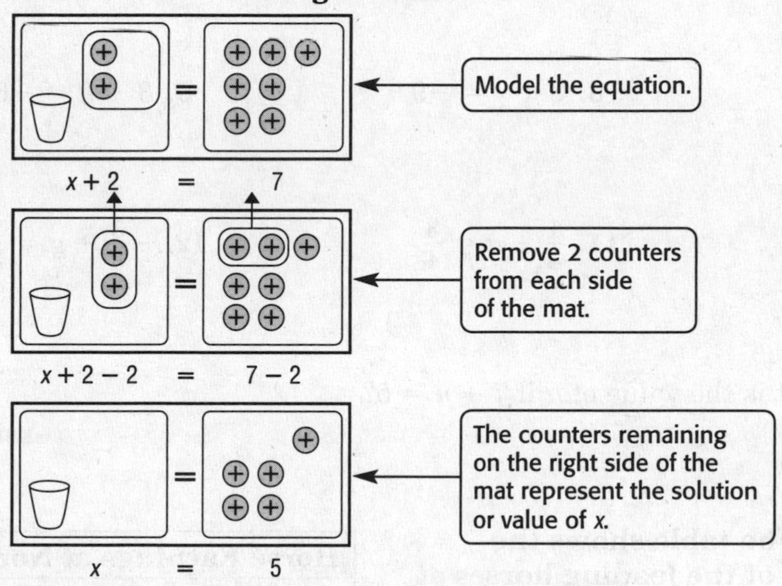

The solution is 5. $5 + 2 = 7$ ✓
5 substituted in the original equation is correct.

Example 2 **Solve $b + 3 = 2$.**

$b + 3 = 2$	Write the equation.
$-3 = -3$	Subtract 3 from each side to undo the addition of 3 on the left.
$b + 0 = -1$	Simplify.
$b = -1$	

The solution is -1.

Check	$b + 3 = 2$	Write the original equation.
	$-1 + 3 \stackrel{?}{=} 2$	Replace b with -1.
	$2 = 2$ ✓	This sentence is true.

Exercises

Solve each equation. Use models if necessary. Check your solution.

1. $a + 1 = 7$ **2.** $3 + b = 8$ **3.** $c + 7 = 4$

4. $9 = x + 4$ **5.** $g + 8 = -2$ **6.** $d + 6 = -5$

Copyright © Glencoe/McGraw-Hill, a division of The McGraw-Hill Companies, Inc.

NAME ______________________ DATE __________ PERIOD ____

12-3 Practice

Solving Addition Equations

Solve each equation. Use models if necessary. Check your solution.

1. $9 + d = -5$

2. $b + 2 = 6$

3. $x + (-4) = 1$

4. $-2 + j = -9$

5. $m + (-4) = 9$

6. $1 = f + (-7)$

7. $6 + c = 3$

8. $8 + y = -9$

9. $3 + h = -6$

10. $p + (-6) = -4$

11. $\frac{1}{4} + a = \frac{3}{4}$

12. $-\frac{3}{8} + g = \frac{2}{8}$

13. **ALGEBRA** What is the value of n if $7 + n = 5$?

THOROUGHBREDS The table shows the earnings of some of the leading horses at Northlands Park. Use the table to answer Exercises 14 and 15.

Horse Earnings at Northlands Park	
Horse	**Earnings**
Sparhawk	$52,800
Griffin's Web	$43,757
Kaylee's Magic	$121,113
Eternal Secrecy	$57,532
Silver Sky	
Huntley's Creek	

14. Sparhawk has earned $8,329 more than Silver Sky. Write and solve an equation to find Silver Sky's earnings.

15. Write and solve an equation to find Huntley's Creek's earnings if the total earnings for all the horses is $354,386.

Copyright © Glencoe/McGraw-Hill, a division of The McGraw-Hill Companies, Inc.

NAME ______________________ DATE ____________ PERIOD ______

12-4 Study Guide and Intervention

Solving Subtraction Equations

Addition Property of Equality If you add the same number to each side of an equation, the two sides remain equal.

$$\begin{array}{r} 5 = 5 \\ +3 = +3 \\ \hline 8 = 8 \end{array}$$

Example 1 **Solve $x - 2 = 1$ using models.**

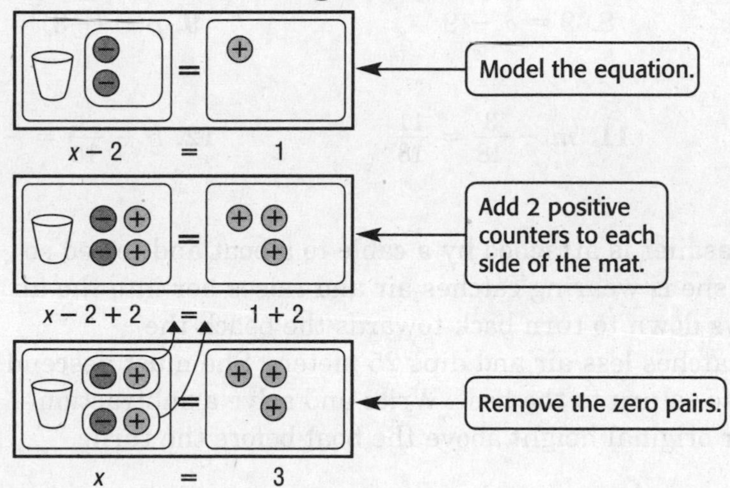

The solution is 3.

Example 2 **Solve $b - 3 = -5$.**

$b - 3 = -5$	Write the equation.
$+3 = +3$	Add 3 to each side to undo the subtraction of 3 on the left.
$b + 0 = -2$	Simplify.
$b = -2$	

Check | | |
|---|---|
| $b - 3 = -5$ | Write the original equation. |
| $-2 - 3 \stackrel{?}{=} -5$ | Replace b with -2. |
| $-5 = -5$ ✓ | This sentence is true. |

Exercises

Solve each equation. Use models if necessary. Check your solution.

1. $a - 2 = 3$ **2.** $b - 1 = 7$ **3.** $c - 4 = 4$

4. $-2 = x - 4$ **5.** $z - 6 = -3$ **6.** $g - 3 = -4$

7. $-9 + w = 1$ **8.** $v - 8 = 5$ **9.** $-7 = y - 5$

10. $u - 3 = -4$ **11.** $-2 = t - 9$ **12.** $f - 6 = -3$

Lesson 12–4

Copyright © Glencoe/McGraw-Hill, a division of The McGraw-Hill Companies, Inc.

NAME ______________________ DATE __________ PERIOD _____

12-4 Practice

Solving Subtraction Equations

Solve each equation. Use models if necessary. Check your solution.

1. $t - 7 = -19$

2. $x - 2 = -5$

3. $g - 6 = -2$

4. $-6 = c - 5$

5. $h - 5 = 4$

6. $8 - j = 5$

7. $y - (-7) = 7$

8. $9 = a - 9$

9. $p - (-3) = 5$

10. $d - 5 = -9$

11. $m - \frac{3}{18} = \frac{11}{18}$

12. $b - \frac{3}{15} = -1$

13. PARASAILING A parasailer is attached by a cable to a boat and towed so that the parachute she is wearing catches air and raises her into the air. When the boat slows down to turn back towards the beach the parasailer's chute catches less air and dips 25 meters. She must descend another 45 meters to return to the boat. Write and solve a subtraction equation to find her original height above the boat before the turn.

14. ALGEBRA What is the value of k if $-6 = 9 - k$?

15. The Petrified Forest National Park in Arizona recently expanded their boundaries by 93,533 acres. The original acreage was 125,000. Write and solve a subtraction equation to find the new acreage of the park.

16. A mako shark caught by a rod and reel in Massachusetts Bay weighed 1,324 pounds. This was 103 pounds more than the International Game Fish Association (IGFA) record. What is the IGFA record for a mako shark?

Copyright © Glencoe/McGraw-Hill, a division of The McGraw-Hill Companies, Inc.

NAME ______________________ DATE __________ PERIOD _____

12-5 Study Guide and Intervention

Solving Multiplication Equations

In a multiplication equation, the number by which a variable is multiplied is called the **coefficient**. In the multiplication equation, $2x = 8$, the coefficient is 2.

Example 1 **Solve $2x = 6$ using models.**

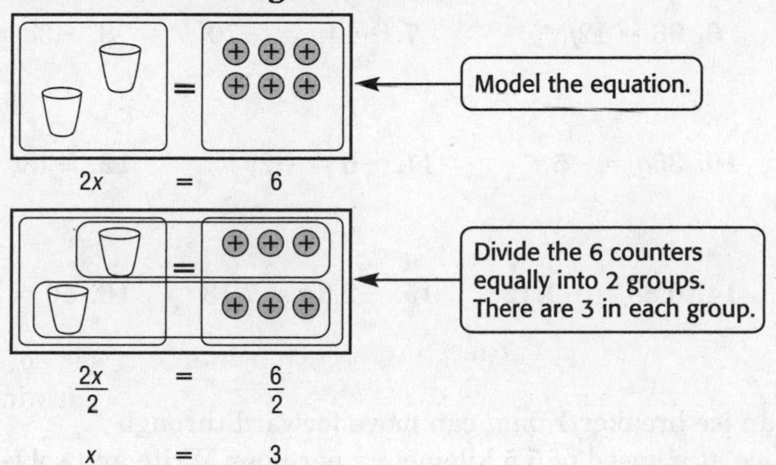

$$2x = 6$$

$$\frac{2x}{2} = \frac{6}{2}$$

$$x = 3$$

Check $2x = 6$ Write the original equation.

$2(3) \stackrel{?}{=} 6$ Replace x with 3.

$6 = 6$ This sentence is true. ✓

The solution is 3.

Example 2 **Solve $-4b = 12$.**

$-4b = 12$ Write the equation.

$\frac{-4b}{-4} = \frac{12}{-4}$ Divide each side by −4 to get a single positive variable by itself.

$1b = -3$ Simplify.

$b = -3$

Check $-4b = 12$ Write the original equation.

$-4(-3) \stackrel{?}{=} 12$ Replace b with −3.

$12 = 12$ This sentence is true. ✓

The solution is −3.

Exercises

Solve each equation. Use models if necessary. Check your solution.

1. $5a = 25$ **2.** $7c = 49$ **3.** $24 = 6d$

4. $2x = -8$ **5.** $18 = -9y$ **6.** $-8g = -16$

7. $18 = -3z$ **8.** $-4w = -36$ **9.** $56 = 7v$

10. $24 = -8f$ **11.** $3u = -27$ **12.** $-42 = 6t$

Lesson 12–5

Copyright © Glencoe/McGraw-Hill, a division of The McGraw-Hill Companies, Inc.

NAME ______________________________ DATE ____________ PERIOD _____

12-5 Practice

Solving Multiplication Equations

Solve each equation. Use models if necessary.

1. $7a = 63$ **2.** $-14k = 0$ **3.** $-13w = 39$ **4.** $55 = -11x$

5. $3v = -42$ **6.** $96 = 12f$ **7.** $-14u = -70$ **8.** $-3c = 3$

9. $15s = -120$ **10.** $35q = -5$ **11.** $-6 = -2y$ **12.** $-13t = -117$

13. $72 = -6r$ **14.** $0.8b = -1.12$ **15.** $-2.3g = 7.13$ **16.** $40 = -1.6m$

14. TIME The Russian ice breaker *Yamal* can move forward through 2.3-meter thick ice at a speed of 5.5 kilometers per hour. Write and solve a multiplication equation to find the number of hours it will take to travel 82.5 kilometers through the ice.

FUNDRAISING A school is raising money by selling calendars for $20 each. Mrs. Hawkins promised a party to whichever of her English classes sold the most calendars over the course of four weeks. Use the table to answer Exercises 15–17.

Mrs. Hawkins' Fundraising Challenge	
Class	**Number of Calendars Sold**
1st Period	60
2nd Period	123
3rd Period	89
4th Period	126

15. Write and solve an equation to show the average number of calendars her 3rd period class sold per week during the four-week challenge.

16. How many calendars did the 1st and 2nd period classes sell on average per week? Write and solve a multiplication equation.

17. What was the average number of calendars sold in a week by all of her classes?

Copyright © Glencoe/McGraw-Hill, a division of The McGraw-Hill Companies, Inc.

12-6 Study Guide and Intervention

Problem-Solving Investigation: Choose the Best Method of Computation

When solving problems, one strategy that is helpful is to *choose the best method of computation.* After reading a problem you can determine if addition, subtraction, multiplication, or division will be the best method to solve the problem. You may often find that there is more than one method you can use to solve a problem.

You can choose the best method of computation, along with the following four-step problem solving plan to solve a problem.

1 Understand – Read and get a general understanding of the problem.

2 Plan – Make a plan to solve the problem and estimate the solution.

3 Solve – Use your plan to solve the problem.

4 Check – Check the reasonableness of your solution.

Example

SHOPPING **At a craft fair, Mia bought a necklace for $6.50, a picture frame for $12.75, and a candle for $4.25. If Mia took $30 to the craft fair, how much money did she have left over?**

Understand You know how much Mia spent for each item. You also know how much money she took to the fair. You need to find the amount she has left over.

Plan One method is to add to find the total amount she spent. Then subtract this amount from $30.

Solve Total Mia spent: $6.25 + $12.75 + $4.50 = $23.50

Amount left over: $30.00 – $23.50 = $6.50

So, Mia had $6.50 left over after she bought the three items at the craft fair.

Check Check the result by adding the amount left over, $6.50, with the amount she spent on each of the three items. Since, $6.25 + $12.75 + $4.50 + $6.50 = $30.00, the answer is correct.

Exercise

SPORTS Jeff Gordon won $1,497,150 as the winner of the 2005 Daytona 500. Sterling Marlin won $300,460 as the winner of the 1995 Daytona 500. About how many times more money did Jeff Gordon win than Sterling Marlin?

Copyright © Glencoe/McGraw-Hill, a division of The McGraw-Hill Companies, Inc.

NAME ______________________ DATE ____________ PERIOD ______

12-6 Practice

Problem-Solving Investigation: Choose the Best Method of Computation

Mixed Problem Solving

Choose the method of computation to solve Exercises 1 and 2. Explain your reasoning.

1. **IMMIGRATION** California registered 291,216 immigrants in 2002. New York registered 114,827 immigrants. About how many times greater was the number of immigrants received in California than in New York?

2. **MONEY** Ingrid bought a book for $1.50 and a CD for $5.50 at a flea market. What was the total amount she spent?

Use any strategy to solve Exercises 3 and 4. Some strategies are shown below.

Problem-Solving Strategies
• Look for a pattern.
• Use a graph.
• Guess and check.

3. **MONEY** Ilia and Candace went to see a movie. Their tickets were $9.50 a piece. They also bought a large popcorn for $4.00 and two medium drinks for $3.25 each. If they have $10.50 left, how much money did they have originally?

4. **PUMPKINS** Mr. Maldonado figures that he sold an average of 26 pumpkins a day. Use the graph to find how many he sold on Friday.

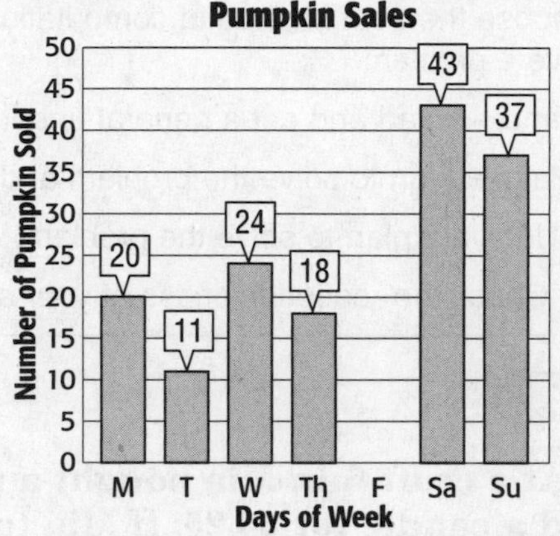

5. **MILEAGE** Mark drove a delivery truck 150 miles on Monday, 63 miles on Tuesday, and 122 miles on Thursday. How many miles did he drive on Wednesday if his average was 106.5 miles per day?

6. **WEATHER** The morning outside temperature is −13ºF. It rises 10º by midafternoon and drops 4º by evening. What is the outside temperature at the end of the day?

Copyright © Glencoe/McGraw-Hill, a division of The McGraw-Hill Companies, Inc.